This is a wonderful story of GI Joes, gardens, and gangsters. Visit Americana, where Berta serves up Hoosier hospitality. It's a timeless journey of family, farms, and forgiveness.

Lori Lehe, Round Grove, IN

Sunset Kings is an unexpected story about the Hoffman farm family in the fictional Indiana county of Belton, who experience the impact war can have for generations. This novel takes you in directions that should satisfy any reader who loves to look back at a time when life was simpler yet more difficult. From the corn fields of Belton County and the Blind Pig speakeasy in Indianapolis to the magnificent French Jardin Botanique of Bailleul, Sunset Kings brings to life a story we rarely read about life on the farm.

Jane Ade Stevens, Indianapolis, IN

This captivating novel begins with WWI, which draws in the reader. The quick transition from war to the farm is very nicely paced. Any farm person reading this will identify with much of the novel, especially the details about what "chores" meant, distances to town, hospitals, etc. It will bring back memories of grandparents, and even parents, talking about what it was like to farm without electricity, using iceboxes for refrigeration, and working with horses and eventually power equipment.

Greg Lamp, Saint Paul, MN

Books by Steven Cain

Sunset Kings © 2020 Upon the Moment Publishing

Library of Congress Control Number: 2021905234

ISBN: 978-1-7368362-0-0

Cover photo: **Kathryn Cain**

Cover Design: **Stephanie Cain**

The Accident in Larson © 2021 Upon the Moment Publishing, LLC

Library of Congress Control Number: 2021923613

ISBN: 978-1-7368362-4-8

Cover photo: **Steven Cain**

Cover Design: **Stephanie Cain**

UPON THE
MOMENT
PUBLISHING

To

Kathryn 'Katt' Cain,

who inspires me.

To

my dad, Stephen L. Cain, and my uncles,

who inspired this story,

Rest in Peace All.

To my readers:

This is not a war novel, although it starts that way in the first chapter. This is a story about what happens to a family after the war. I hope you enjoy visiting the Hoffman's.

I wish a special thanks to the following for

their help:

Kathryn 'Katt' Cain who helped every step of the way

Stephanie Cain for design creation and consulting

Tracy Petersen for carefully editing the novel

Randy Spears, novelist, who gave me great advice

Rosie Lerner for background help on the Jardin Botanique

Steve Doyle, publisher, who pointed me in the right direction

Stuart Boehning, for setting up Upon The Moment Publishing LLC

For reviewing the novel

Lori Lehe

Jane Ade Stevens

Greg Lamp

My professors at Purdue University for helping me understand novel writing and better understand agriculture

Chapter 1:

Summer 1918

In the summer of 1918, in the French village of Bailleul, just a few miles from the Belgian border, the Allies advanced. They were attempting to push the Germans out of France and Belgium. Because of his excellent German and serviceable French, John Hoffman joined an advanced squad of British, French, Indian, and Belgian fighters in observing the front and communicating back to the infantry and artillery sections. As scouts, they reported the locations of the German headquarters and the placement of troops. A perilous job, but one that fulfilled John's desire to free this region of France and ultimately the German people from an oppressive dictatorship. John had read about Kaiser Wilhelm II and the leadership of Field Marshall Paul von Hindenburg and General Erich Ludendorff before he joined the battle. He hated them for what they did to this region of France.

John's European heritage helped him easily assimilate into the squad, unlike the many thousands of his fellow American soldiers who were there to push their way into war-torn villages. John was a strapping man, aged 34, chiseled out of the Hoosier prairie and a family cattle farm. Fair of hair and eyes, his blond hair was cropped so close, his ears appeared to stick straight out from his head. Yet, his straight, strong Roman nose helped his ears balance on his face. His complexion was dark, especially in summer. His brow, when furrowed, appeared to extend over his eyes, often giving him a look of constant scrutiny. It was fitting because John was calculating and methodical in all that he did.

John had traveled to this region of France, where his ancestors thrived for centuries, in the aught years of the 1900s, when he was in his early twenties. He was a first-generation American, born after his father, John Hoffman I, arrived in America from a region of France where the residents spoke both German and French dating back to the Holy Roman Empire.

John fondly recalled visiting Bailleul with quaint houses, cathedrals, and immaculate French gardens. His great uncle, Pere Hoffman, in his late years, had volunteered to help maintain the Jardin Botanique of Bailleul. Pere was ecstatic to show it off to John during what turned out to be the last summer of his life.

Bailleul had suffered at the hands of the Germans from the beginning of the war. Now, as the squad lay quietly in the rubble of the village, John thought about his visit to the garden with his great uncle.

He vividly recalled the oval-shaped hedges in a sea of red roses, surrounded by neat rectangular hedges. He remembered walking through the garden and sampling the smells of the Jardin Botanique. The deep, red-rose scent had permeated his mind as well as his nose. He was surprised by the apple aroma of some peach-colored flowers; his uncle said they were Lady of Shalott roses. Pere Hoffman smiled and, with one finger across his lips, said, "Don't tell anyone. We're not supposed to have English roses in our French Jardin, but I've always thought that was simply a guideline." The scent reminded John of the apple strudel that his mother baked with fresh apples from their farm in Indiana.

Pere Hoffman shined with pride as John was obviously impressed with the Jardin. A tall fountain with a statue of a nymph stood in the center. Pere said it represented the "giver of spring rains." At each corner of the garden stood ten-foot-tall satyrs, all slightly shorter than the nymph. Though he had no military experience, John's uncle said presciently, "We hope that the delivery of rain is more important than the defense of the Jardin." He had no idea what was soon to come.

These satyrs featured horse-like ears and tails, sharply chinned faces and deep-set eyes, long necks, and muscular abdomens. John easily imagined these exquisitely carved creatures might turn their faces to scope him, the intruder, to determine if he was a threat. Their brows and cheeks were chiseled into mean, forward-thrusting faces as if to protect the Jardin. Their penises were erect to represent the garden's fertility.

John blushed when he saw this, but his uncle scolded him. "Why the embarrassment? Fertility in any form is a God-given privilege and one that we shall not take in vain, but rather in a matter of rejoicing in life." Still, a little embarrassed, John admitted that he appreciated the French portrayal of fertility.

Suddenly, a rifle shot burst and whistled above John's head. The reality pulled him back from his reverie with a force that spun his mind. His sensual traverse from apple strudel to mud, piss, pus, and sweat sped by in a second. More shots were fired, and John's squad fired back. They had a job to do. Observe the German hideout in Bailleul and report their findings to the infantry and artillery troops.

Fortunately, wireless communication had become transportable and effective during the war. A mobile squad such as John's could carry fieldsets and communicate with command. Taking heavy fire, the men acted quickly. An Englishman in the squad, Frank Trevelyan was a communication expert and John's steadfast accomplice in reporting on the Germans. Frank had the appearance of a heavy man who had lost weight during the war. When not in battle, gleeful green eyes peered from his face. His brown hair was shaved so close to his head that he appeared bald when wearing his helmet. Another Englishman, Barry Bernard, was a tiny man and a maps expert. The men knew their jobs.

Frank and Barry had met in the Battle of Ardennes. In February 1914, they were among the first British to serve with the French during a brutal winter. They sought each other out to reminisce about their homeland. They immediately caught on to each other and offered support when it was needed. The transition from everyday British life to a stagnated battle, where more than a million soldiers died or were wounded, warped their minds. They needed each other to hold onto the notion of who they were and what it was to be British. They had grown up under very different circumstances in England. Frank's parents were wealthy and traveled extensively, taking Frank along with them. Barry's father was a successful butcher, but his mother ruled the roost. Because the two men had such different backgrounds, they had plenty to talk about as they shared the stories of their lives. The one thing they had in common was that they were British, and the enemy wasn't.

A year before this new engagement at Bailleul, they had fought together in the Battle of Passchendaele, a Belgian village not far from their present location. In that battle, they had slogged through flooded mud fields born from unusually heavy rains and wrecked drainage. Tanks were nearly rendered useless. Many soldiers on both sides drowned in trenches, bogs, and mud holes. After months of fighting and nearly a million casualties, the British overcame the Germans and gained Passchendaele. This eventually led to a special mission to scout the enemy in Bailleul, where they were that night.

Frank and Barry's strong camaraderie wasn't unique in war. What was unusual was that they had developed a quick method of communication in battle, one that those around them could never comprehend. The ability to communicate with few spoken words was a gift when scouting behind enemy lines. Frank was a natural with field communications, and Barry took to maps and geography like a kid to a playground. Even when they sat in a trench away from the action, Barry studied maps like a leopard studies its prey.

Tonight, under fire and now glowing flares, the unit moved into position to report on the Germans. John gasped at the first brief flare. Before him stood the Jardin Botanique. Though nearly all of Bailleul was reduced to a shamble of brick and lonely walls by the German bombings of 1914, there stood the Jardin that his great uncle had shown him ten years earlier. Beyond the Jardin lay a street and then a three-story, former administration building that provided a headquarters and cover for the Germans.

For a split second, John wondered if even the enemy thought this garden was too precious to destroy, or maybe it was just an inconsequential space that had escaped the war.

Frank and Barry had closed in on the wall to the Jardin. Around them, the stench of the bloodied earth and rotted bodies of the Allies would make most normal people vomit. The Germans had removed the bodies of their fallen soldiers, but not those of their enemies.

"This rotten air would make a surgeon toss his chunks," Frank said. That would not happen to these men. Though they could not ignore the stench, the soldiers had tragically become so accustomed to it that they proceeded as if they were on a football field practicing their next play.

Barry studied his map and began to write down coordinates that Frank would radio back to command. As quick as a snap of lightning and just as deadly, a German bullet pierced Barry's helmet and skull as he extended his neck to observe the lay of the land. He fell limp upon his map. Frank, still taking cover, grabbed Barry's shoulder.

"Barry, Barry!" Frank shouted. "Talk to me."

John, who had a better view of what had happened, crawled over the rubble, trying to ignore the sharp pain of the stones against his elbows and knees. He turned Barry over to view the map. He glanced at Frank but had no time to console him. The coordinates that Barry had written must be communicated. What John saw terrified him. Barry's blood stained the map, making it nearly impossible to read Barry's writing. Frank, whose battle experience overshadowed John's, quickly steadied himself. He prepared his radio while others in the squad returned the Germans' fire and kept them at bay.

"What are the coordinates?" Frank shouted.

John wiped the sweat from his brow to help clear his vision. "I'm not sure," he stammered. "The blood...the blood has stained it badly."

Frank shouted again. "What are the coordinates? We must wipe out these dirty bastards."

John made his best guess, and Frank, without hesitation, radioed the artillery division but begged them to wait ten minutes before firing. Abhishek Singh, the squad's non-commissioned officer, immediately ordered the men to retreat. Singh was the first Indian soldier awarded the Victoria Cross for efforts in a battle near Ypres, Belgium. The Germans ceased firing after the Allied squad retreated. With the gun fire from both sides fading, the squad crawled away, and when they could, they ran. In that ten minutes, Singh's men covered more than a mile. John wondered if the Germans were praying; he thought he heard slow, solemn singing, but he wasn't sure if that was just his ears ringing. Surely the Germans knew what the cease-fire indicated. Exactly ten minutes after Frank's report, Bailleul glowed, and with the crackling of timbers, it sounded like a world on fire.

Returning to field headquarters, the men walked quietly through a long-neglected field. The soft grasses and weeds were a soothing contrast to the piles of rubble in Bailleul. Alone and within themselves, their minds raced. Frank reflected on Barry, a wartime gift, ripped from him by one German bullet after the millions of bullets they dodged in the last four years. John, relatively new to the front line, viewed the dangers differently. He had passed through Brest, a port city in France, where influenza had killed thousands of American and Allied soldiers and left tens of thousands more suffering from the disease. What little news there was about the influenza came from Spain. John had felt as if the Allied commanders kept information about it from the troops, which was worse because it was all around them. John was content outside Bailleul, away from Brest, where he could make a difference.

Now far from Brest, John still thought of the port city's smell and the fear in that city. He could not help thinking about Brest, far away from the front lines, the fear in Brest's military and medical camps, and the sense of doom that far surpassed those same feelings at the front line, if only for John. He was a doer. Here, a soldier could act and could make a difference. In the medical camp, John had watched some of his closest military friends lie helplessly on cots. The puke and feces stench, at the peak, was too much for those cleaning the camp, punctuated the fear of death, filled the air. Fear impaled the eyes of soldiers so far from home, with so little to help them fight the flu. Malnutrition didn't help. John thought it was a major contributor to the flu-related deaths.

But this was not Brest. It was the front lines, where time and time again, John had helped the Allies identify the retreating enemies' strongholds. He had helped in that effort tonight and would walk back to the field command to take on his next assignment. Yet, one thing burned in his mind. The bloodstains over the markings that Barry had written down. It was a fine line between the Jardin and the administration building. Did he get it correct? This would gnaw at his mind.

The gray-blue dawn recaptured the cloudy sky when the squad returned to the field command. The crow of a rooster comforted John as they reported back. The field commander received their oral reports, while a British stenographer captured their reports for historical records and future planning. The commander gave the squad their next assignment. They accepted it by doing nothing but listening to the assignment's details, and they waited to hear who would replace Barry. The commander pointed to his maps, and John could clearly see that they would not pass through Bailleul. He was sorely disappointed. The commander asked for questions, and John didn't hesitate.

"What of Bailleul?" he asked.

"Your job is done there," the commander snapped. "Already this morning, aerial recon confirmed you succeeded in your mission. Other troops will cover Bailleul. You are needed elsewhere." The commander turned to Frank and John as an afterthought and quickly said, "Good work! Now go get a much-needed meal and rest...if you can."

The others turned, but Frank hesitated. "Sir," was all he said before the commander interrupted.

"Your report was noted," he stated brusquely. "Your man will be carried out, taken care of, and family will be notified."

Not nearly satisfied, Frank turned and walked out of the tent because it was his only option.

Chapter 2

April 1919

John stood on the deck of a ship on a chilly spring day, salty ocean air mixed with the grimy, oily odor of the ship's burners and engines greeting his nostrils. It was April 1919, and John was finally released to go home. Nervous with anticipation and guilt from being apart from the boys at the start of the spring season on the farm, he began the voyage across the Atlantic.

The crossing took seven days, a day longer than usual because this was the crew's first voyage on the ship. This vessel, the Graf Waldersee, was a captured German troop transport ship, and the intricacies of the German boat slowed down everyone from the commander to the coal crew members. After returning this set of troops to New York City harbors, the ship would be converted, at the cost of $160,000, to be a more efficient troop transport. Ships sailing under flags of many nations were being used to bring home the massive numbers of men who had fought overseas. The irony that John was on this ship didn't escape him because he was one of the few U.S. soldiers who could read the German signs, which had yet to be converted to English. He wondered if he was put on this ship for that reason, but he resigned that thought as too much thinking on the part of the dispatchers.

Because John could read the signage, he felt more comfortable aboard the ship than most of the other men. Some were outright scared, and there were rumors that they might be attacked by Allied ships. When one man voiced that concern, John simply raised his right hand and proudly made a sweeping arch over his head, indicating the plethora of U.S. flags that flew above them. Others took matters into their own hands, in some places crossing out the German words with dark XXs and replacing the words in English. The soldiers joked among themselves that they could understand what the word "toilette" meant because of its similarity to the English word 'toilet' as well as the smell. They left that word alone.

The ship was exactly twenty years old when John boarded it. Originally a passenger-cargo steamer named after Alfred Ludwig Heinrich Karl Graf von Waldersee, a German field marshal who had passed away several years before this war, the vessel eventually carried 5,600 war veterans from Brest to New York. It contained only first- and second-class sections, increasing John's odds of securing a better cabin than if he'd sailed aboard a ship with a third-class section.

John jumped at the chance to sail home on this ship because it departed several days ahead of the others also headed to the United States. As John approached the ship at the Brest docks, he knew he would never forget this moment, the American flags flying from each of the four masts of the former German vessel. Along with his fellow troops, John marched up the ramp like he was in a polka street dance, wafting back and forth. Instead of music, the clanking of machines rattled John's ears. The ship bore a single funnel from which dark smoke would billow across the Atlantic. Graf Waldersee's beam spanned 53 feet, a draft of almost 24 feet, and it was powered by a pair of steam engines.

John knew the ship would be crowded, with as many bunks as possible crammed into each room. He wondered how many men would be seasick. That certainly would be a smelly problem to endure on a seven-day trip. He had been told no draft animals would be shipped on this one because it wasn't set up for them. The excitement of returning home glowed from the soldier's eyes. In a weeks' time, they would arrive in New York City.

John's thoughts returned to the farm and his wife, Berta. In the mess hall, he thought of the song, "How Ya' Gonna Keep 'em Down on the Farm," sung by a soldier, a fellow farmer. Songwriters and playwrights alike had contemplated how rube farm boys, turned into men who saw the war and urban Europe, could be happy with farm life after they had seen "Paree." John hated the idea of the song.

He had seen the celebration in "Paree" after the end of the great war. He usually observed from a distance, drinking alone while other soldiers drank and took advantage of drunk Parisian women's desires to thank them. Quite by accident, he ordered champagne. It was his first taste of the bubbly drink, which surprised him. He asked the bartender in this English-friendly bar what it was called.

"Is this wine okay?" he shouted at the bartender over the loudly singing soldiers and their acquaintances.

"Ah, monsieur," the French bartender said with a smile.

"Perhaps you've never had champagne, eh?" John shook his head. The bartender replied, "Well, enjoy a French novelty, my friend." John did. He had never tasted anything so delicious. He ordered an entire bottle and retreated alone to the back of the bar. He made plans for his return as he sat: recouple with Berta, hug his sons, and run the cattle operation and farm.

Again, thinking back to the song and keeping the boys down on the farm, John realized it was condescending in every way. He knew many, many soldiers would be drawn to the mythical appeal of city life. In fact, by 1920, urban dwellers would narrowly surpass the farm population for the first time. The year 1920 would mark the beginning of a new America. In the back of the bar, John readied for it. He wouldn't go to the city. He knew he would run back to the farm, retreating from the new way of life, as quickly as possible. He sensed that hard times would come, and he felt the farm would be the best place to be when they did.

John had learned about economics from the ag club programs of Purdue University, Indiana's land grant school. Purdue's outreach began just a few years before John went off to war and had prepared him for the changing economy. Purdue professors had ridden trains to local stations and talked about everything from agronomy to agricultural economics. In 1916, the year before John left for Europe, he had had a chance to talk extensively with a professor over coffee in the train station about the farm economy. This conversation, along with his keen sense of how things worked, prepared John for what might happen next. It disturbed him to his bones. The U.S. government encouraged overexpansion of agriculture during the war, and John thought this could only lead to overproduction when the fighting ended.

After arriving in New York, the army sent John to Fort Slocum, a military post on Long Island that had been used to process soldiers since the Civil War. The soldiers slept in tents, endured physical inspections, and waited for their discharge papers. John was thankful that the April weather on Long Island was milder than usual for the two weeks that he waited there. As soon as John was processed out of the army, he bought a $38 ticket on the 20th Century Limited that took him from Grand Central Station in New York to LaSalle Station in Chicago. The ticket cost an extra $9, but this one got him home in twenty hours, four hours faster than a normal train.

John and his wife Berta had planned to meet at the station in Chicago, but she wasn't sure when he would arrive. The mail was a sloth compared to the telegraph, so she drove into nearby Lafayette, Indiana, each day for the two weeks in hopes of receiving a telegraph with John's arrival date and time in Chicago. The day finally arrived, and she immediately drove north to LaSalle Station, where they previously had agreed to reunite.

Berta was waiting when John's train arrived. He spotted her first through the window. Today, Berta, a strong, radiant, thirty-four-year-old woman, sparkled even more in John's eyes. She smiled widely at the thought of reuniting with her husband, the rosiness of her cheeks accentuating her porcelain skin. Berta wore her reddish, gold hair short; her curls lightly touched the nape of her long slim neck, which she knew John adored. She wore a shimmering red and gold speckled dress that complemented her complexion. The dress was cut low, and piping crossed over her bosom.

John and Berta gasped when they laid eyes upon each other, and their faces shone with joy. They embraced on the train platform, their reunion more passionate than they expected. The joy of being reunited, coupled with the relinquishing of the fear that that might not have been possible, led the two to shed their typical shyness for public displays of affection. They held each other for minutes. Berta smelled heavenly, all the sweeter for John after so many years at war. On the contrary, John smelled like a long-distance traveler, but Berta didn't mind. After a few moments, John said, "Let's get out of this station."

Berta, the more experienced driver in downtown Chicago, drove to the Auditorium Hotel, a large hotel with a lavish auditorium and a view of Lake Michigan across the street. No strangers to Chicago, they chose the hotel because it was large, and they were assured of a room for the night. Berta had suggested the idea of one night in Chicago, and she was intrigued with this hotel. Across the street from the hotel was an open park with a backdrop of Lake Michigan. The view impressed the couple, who strolled in the park before retreating to their room. John was grateful to stretch his legs after the long train ride, and the two gulped in the fresh smell of Lake Michigan.

Walking hand in hand through the lakeside park near their hotel, the couple anticipated the night to come. They barely closed the hotel room door before they pawed at each other, loosening buttons and passionately kissing. The room held a different perspective for each of them. Berta became bedazzled by the city lights outside, and John loved the comfort of a real bed and the beauty of his wife by his side. Their passion for each other led to its natural end, again and again for a couple of hours.

The next day, the couple bounced along the rough road in the Model T that Berta had driven to meet John. They were surprised by Chicago's sprawl along Highway 41 into northwest Indiana; Europe's growing demand for U.S. steel fueled a boom in the Hoosier steel industry. It took four hours to drive home to the Hoffman farm near Boulton, Indiana. The setting sun shone brightly on the cattle farm, and John swelled with pride at the view. He watched as they drove past the strong, red-coated Gelbvieh cattle that John had missed so much during the war.

Berta smiled when she watched John gaze over the cattle, reading his mind.

"Calving went very well the last two years," she told him. "We did very well at the Indianapolis market."

The pasture and farm air smelled fresh and rich to John, who had suffered the odors of battles for two years. John's face turned from a bright smile to a doleful expression.

"Did I say something wrong?" Berta asked.

"It's not you," John replied. "It's the change that's coming. We did well in the markets when half the world needed our commodities." John had had plenty of time to think about the world's economy during his voyage home. Europe's countries and other war-torn areas would return to productivity. That, in turn, would burden the American farm commodities market, which he feared had overexpanded. He had read that agricultural debt skyrocketed as farmers adopted the latest technology during the war years. He thought about future farm struggles as commodity markets shrunk.

John willed away these thoughts and returned his gaze to the cattle farm. He hated the term "rancher," even though his farm was large enough to be considered a ranch. He embraced the word "farmer," a person who actually worked on the land rather than just managed an operation. John could not wait to get back to caring for the cattle and tending to the crops.

Gelbvieh cattle reigned in this German area of Indiana. They were bred from several cattle breeds in the Franconian districts of France near Alsace, where John's family farmed for decades. John's grandfather and father brought the cattle with them from France to the rich pasture lands of Indiana. His grandfather and father did well, even during the ups and downs of the farm economies of the late nineteenth century. Once in America, their adherence to conservative farm growth had given them a solid foundation. Still, their real financial advantage came from selling their land in continuously overcrowded eastern France and investing it in very productive farmland in Indiana. John's father had scouted this part of the state. He bought the farm as part of the land-grant sales by the U.S. Government. He paid one dollar per acre. Though the Indiana skies were more open than in the hilly country of France, the prairie grass grew lusciously.

Underground water stayed just a few feet below the surface, making windmill-powered pumps an easy alternative to water the cattle. Swamps still covered some of this land, which meant a plentiful water supply.

While other Indiana farmers favored breeds like the sleek black Angus or the often-red Hereford, which were brought to the United States almost one hundred years ago, the Hoffman's preferred the Gelbvieh. They knew how to manage these multiple-purpose animals. They not only provided meat and milk, but they stepped in as draught animals. By 1900, John emphasized meat production, taking only enough milk to provide for him and his neighbors who no longer milked cows. Financially, John appreciated the breed's ability to gain weight efficiently. Gelbvieh females produced calves that matured quickly and gave the calves a better chance of survival during the harsh Indiana winter months. For esthetic reasons, he enjoyed the red color that contrasted with the green pastures on his farm. The young cattle looked like deer standing out in the pasture.

John's reverie was cut short when they pulled into the lane, and he saw his boys. His sons Karl and Fritz stood on the driveway between the house and a big red barn. They had stayed home on the farm to tend to the daily chores while their mother went to

Chicago. With Berta's help, who had been raised on a farm, Karl had taken on the management of the Hoffman family operation at the early age of fifteen when John left for the war. Karl was too young to fight but mature enough for the day-to-day farm work. To the surprise of their gossiping neighbors, Berta and Karl made a perfect management team, each of them bringing unique skills to the job. After the war, the mother and son struggled with the idea of relinquishing management. Farming had changed in the years that John was away. They knew John would need to update his skills before he could return to manage the farm. As he returned to the farm, no one had any idea what would motivate John after the war.

Karl was now seventeen, going on eighteen. John was ecstatic that the end of the war meant his sons would not be drafted. Whereas Karl was dutiful about his farm chores, fifteen-year-old Fritz was a different story. He had his mind set on a life beyond the farm, but because of his father's absence during the war, he dutifully tended to his chores even though he did not share his brother's enthusiasm. John leaped from behind the wheel of the Model T to embrace his boys, a tear of sorrow of lost time mixed with joy rolled down his cheek. He couldn't help but marvel at what fine young men they had become in his absence.

"You've kept it going," he said to his sons. "It looks very good." He stretched his arm out and pulled the boys toward him.

Those years that John was away had been hard on Karl, but he grew in proportion to the task. For Fritz, those years had been hell. He didn't speak of it because he knew it couldn't compare to what John had experienced in the war, but the farm labor still changed him, and not necessarily for the better.

"You're just in time," Karl said. "Amelia has supper ready. We just checked."

Although he was weary from travel, John looked forward to dinner with his wife, boys, and Amelia, who always dined with the Hoffmans. Practically a family member, she felt comfortable with them. In fact, later, John began to notice something in Karl's face when he looked at Amelia, often smiling at her.

John grabbed his bag from the back of the Model T and reached for Berta's hand as he looked upon his home. His father had built it in the Queen Anne style in the late 1890s. The home's turret and gables made it one of the most outstanding in Belton County, and it suited a large cattle farm. John stepped up onto the porch, which wrapped around three sides of the house. Large columns supported the porch roof, and five dark mahogany rockers graced the porch on the west side of the house, the best vantage point to watch a sunset.

John entered his home for the first time in years. Amelia, a buxomy woman with long dark hair contrasting with her pale skin, waited in the kitchen, filled with the delicious aroma of breads, meats, cheeses, and relish. John kissed her on the forehead and rather loudly exclaimed, "At last, again, a home-cooked meal! Amelia, it is very nice to finally meet you." During the war, Berta had written to John, explaining that she hired Amelia to help with the household chores so she could work with Karl and Fritz on the farm. Amelia stood next to the white porcelain-enameled steel stove with light blue trim. A wooden box next to the stove held the wood used to heat it. John looked around the kitchen. A wooden shelf with hooks for coats held green jars full of seeds and spices.

"Let's eat," Berta said, and the family and Amelia sat and talked about American life. It was all still new for Amelia, who fled Bavaria before the war. They talked about changes in the United States during the war, American rural life, and of course, politics, but eventually, the subject turned to the war itself.

"What was it like to fight in the trenches over there?" Karl asked. "I'm sure the newspapers didn't get it right." He glanced at Fritz, knowing how his brother loved reading newspapers.

John grew silent. The innocent question instantly transported him back to the days of military bombardment, a hundred thousand dead over a small land area. He thought of the stench, the open screaming, and crying. John finally looked up, unaware of how long he had been alone with his sour-smelling memories of the battles. All were silent and wide-eyed as they awaited his answer.

John looked down again and jabbed at his food with a fork. All he could muster was, "Not now."

Berta touched his arm, and John automatically withdrew a post-battle reaction. He made himself warm to her touch and put his hand on her arm. He looked around the room at his family and said, "Just not now."

Even John did not know when he would be ready to talk about the war. To tell his loved ones what he had experienced proved to be very difficult. John could shoot an enemy and help radio coordinates that killed many men but talking to his family was too painful. It was an admission that he had grown into something about which they knew nothing. At that moment, Amelia began collecting the empty dishes, and Karl was quick to help.

Berta and John retreated to their bedroom, where the moon-filled night shone through their window. They held each other and talked about the future. Fritz moved into the family room and read a three-day-old Chicago City Times that he had purchased that morning when he took the horse and carriage to the country store known as The Station. Fritz was a loyal reader of the City Times, even though the paper was two or three days old when he got his hands on it. He loved to read about Chicago's busy city life, which seemed worlds away. Despite the distance, by reading the newspaper, Fritz was gleaning details about the world around him and absorbing the vocabulary of the changing landscape. He noted to ask his father when he was ready to discuss the war if he used the nickname "Jerry," a slang term for Germans that he had learned about in the newspaper. In the meantime, as Fritz read, he heard Amelia and Karl clinking dishes and laughing.

Chapter 3

May 1919

Within a few weeks of John's return home, Berta knew she was pregnant again. It was late May, and the sun shone brightly into their bedroom with golden rays. She folded the morning's laundry when the feeling struck; she barely reached the chamber pot under her bed before she vomited. Wiping her lips with her hands, she sat on the floor. A smile grew on her face. Would this one be a girl, she wondered dreamily. John was outside with the boys doing chores. She would tell him tonight. In the meantime, she had a chamber pot to clean.

Not much of a socializer, John usually stayed on the farm rather than running into town, but more and more, he began going to The Station to play euchre and drink moonshine with others. If he needed something from Boulton, he'd send Fritz, who loved going to town. Although he was a farm child, Fritz looked and dressed like a businessman from the city. He kept his brown hair cut short and combed perfectly. His dark eyebrows and straight lips focused his gaze. He gave the impression that he always knew what he was doing, which wasn't always the case.

There was no new word for the day this morning, but something was disturbing in the news. Fritz read about a small development in southern Indiana that shook his core. The article explained that a group of Protestants sought to revitalize the Ku Klux Klan and noted the dangers of the possible expansion of the organization. Fritz felt ashamed for his Hoosier state. He thought, how could this reoccur after the evil days following the Civil War? A so-called leader of the KKK was quoted as saying their goal was to keep Indiana one hundred percent American. Fritz knew that this movement, if it grew, would play hell for blacks and German Catholics in a state dominated by white Protestants, who now considered themselves native Hoosiers.

Fritz didn't know how to process what he had read. At dinner that night, he decided to keep that information to himself and instead talked about more new words that he had read in City Times papers. "Did you know there's already been 112 new words in the American language this year?" He looked around the dinner table for a reaction. Everyone acknowledged him. They also gave him a look that said," Okay, what's the new word of the day?" Fritz smiled and said, "Well, this one was actually from a paper last week. It's the word 'Jerry.'" Berta looked puzzled. Karl moved his hand in a motion as if to say, 'Come on, what is it?' John, familiar with the nickname, looked down and played with his food. Fritz continued, "It's a new nickname for Germans, especially German soldiers. You sure put those Jerrys in their place, didn't you, Pop?"

John barely raised his head. He wanted to put all of that behind him like a guilty dog hides a chewed-up shoe. Even so, he never censored his family. The thought of fighting for their freedom of speech somehow allowed him to relieve some pain of the war.

He didn't answer Fritz but said, "I hope you boys get some good rest tonight. We've got a lot of work to do tomorrow to prepare for calving." At that, John stood, left the house, and went to the barn. Karl looked at Fritz and said, "Well, look what you did. I think we've learned that he doesn't want to talk about it." In the barn, John opened a bottle of moonshine and took a big drink.

John figured the Gelbvieh would start caving that week, which meant they needed a great deal of attention. He didn't stay in the barn for long. When he returned to the house, he sat on the porch and waited for the house to cool. Berta joined him quickly while Fritz stayed inside and read the City News, and Karl helped Amelia with the dishes. When Berta told John that she was pregnant, he reached out and held her hand; a slight smile grew on his lips.

That night, John couldn't sleep. He climbed out of bed a couple of times to check on the cattle. Each time, he carried a kerosene lamp for light. One had to adjust these very carefully because they would coat the lamp lens with black smoke and reduce the amount of light the lamp shed. Also, he walked very carefully with the lamp because when he held it in the wrong position, it blinded him more than helped him. On this trip to the barn, John thought about Berta's news. Of course, he was happy. He remembered sitting on the porch, holding Berta's hand as she expressed her desire for a daughter. Still, darkness clouded his thoughts about his abilities to be a good father. The war had taken a toll that he would pay for in the future. With each trip to the barn, he most assuredly had a few nips of moonshine. Looking at his supply, he saw that he was running out.

The next morning, he and the boys prepared huts for calving. Visitors unexpectedly appeared in the driveway at least once a week. Some would come to drink Berta's coffee with John and the boys. Today's visitors turned out to be Bob "Moonshine" Moore, a neighboring farmer, and his hired hand, Brian Alan Ryan. Fritz nicknamed him "Bar" because of his initials, the funny sound of three first names, and because Bar had become the leading supplier of moonshine. Bar liked the nickname because it was simple. Bob was the opposite of John in every way. A devout Irish Catholic, he farmed a much smaller acreage as a tenant. A tenant's profits were slim, but Moonshine found other ways to make ends meet. Today's visit was more than social.

Berta watched from the turret as Moonshine and Bar drove up the driveway. They bounced with each mudhole in their family flivver, a converted Model T that Moonshine had bought from John in 1915. John had converted the car in 1911, cutting off the backside of the cabin of the Model T and installing a large metal box on the frame. The result formed a farm utility vehicle that was years ahead of the brilliant minds at Ford, Chevy, and Dodge. It was ideal for hauling anything from pigs to, well, moonshine. That's what Bob and Bar hauled today. Now eight years old and not well maintained, the vehicle showed its wear and tear. John wondered how the bucket of bolts still functioned.

As the men approached, John told the boys to keep working while he went to greet the two moonshiners. He smiled and reached out a hand as Moonshine stepped out of the rig. He nodded at Bar, on the other side of the vehicle, in recognition. Bar's tall cowboy-style hat made him look six and a half feet tall, but John figured him about six foot two. His rugged, Irish face and thick chin gave him the look of a ruffian. On the other hand, Moonshine had the appearance of a businessman rather than a farmer, with a long face, neatly trimmed hair, and a scally cap fitted neatly on his head. Moonshine grabbed the stogie from his mouth and pointed it toward the field.

"Looks like you'll have a good hay crop, and it'll be ready for first harvest soon," he said. Moonshine and the other neighbors respected the Hoffman family's farming abilities and were impressed with how Berta and the boys had kept up the farm during the war.

John nodded. He couldn't wait to harvest the first hay crop since his return. The men talked more about farming for a few minutes before Moonshine said, "Let's look at what I brought ya'."

As they walked to the back of the vehicle, Berta watched with disapproval. She knew what they were up to and didn't like it. She would address John later that night when he went to bed with the smell of moonshine on his breath. She knew that her appeal to ask him to stop drinking would fall short of convincing him that he was on the wrong path. John was drinking more, much more, than before the war, but she wasn't sure what to do or say about it. Like thousands of other women at the time, she endured the drinking and hoped that prohibition would come sooner rather than later. Fritz kept the family apprised of the political news, which indicated that women would soon gain the right to vote in Indiana. That would allow them to push for prohibition and more ways to rid the country of alcohol. More than ever, Berta fervently hated drinking because of what it was doing to John. He was never mean or malevolent to her in any way. Others, like Moonshine's wife, suffered abuse from their drunken husbands. She and John loved each other deeply, and she hoped it would never come to that. His drinking put a distance between John and Berta, a distance that left Berta out of John's love for her. Many nights he fell asleep in his recliner and didn't come to bed until after midnight.

John handed Moonshine two twenty-dollar bills, lifted a crate with six gallons of moonshine, and hauled it into the barn, where he stored it out of Berta's watchful eyes. When he reappeared, Moonshine asked, "Berta isn't giving you any shit about drinking, is she?"

John gave him a look and said, "Let's just leave that right there, Moonshine." Bar smiled nervously because he didn't like the confrontational questions that Moonshine often put forth.

Moonshine continued, "It's just that these women need to be put in their place, and you are a decorated soldier in the greatest war to end all wars...." His comment trailed off.

John changed the subject. "How many more stops?"

"You're the first, and you got the best," he replied. "I've got five more stops. So, I might as well get along on my business." Moonshine was neither angered nor embarrassed by John's reaction. He was just doing his business. The men shook hands again, then Moonshine and Bar climbed into the vehicle and drove away. Fritz would tell his family that the men left in a jalopy, one of his newest words. He explained that the longshoremen in New Orleans referred to scrapped autos as jalopies because they were destined for scrapyards in Jalapa, Mexico.

John watched the moonshiners drive away, then glanced at the turret window. Berta was no longer there. He returned to the barn and, with a smile, opened a bottle and took a good, long swig. It was the beginning of an end for him as he continuously tried to reckon with what he had done in Europe. He took another swig and then returned to help the boys set up for calving.

Karl could smell the booze on his dad's breath. "Any left for me?" he asked. At seventeen years old, Karl looked stronger than any man in the county, a product of throwing hay on wagons and wrestling cattle. Karl's dark skin and angular face gave him a gladiator appearance like his father.

John looked at Karl nonplussed. The idea of his boys drinking had never crossed his mind. They were so young when he had left for the war. He smiled and pointed over his shoulder. "It's in the barn on the second shelf. Go and get back here. We have a lot of work to do." Fritz dropped his pitchfork and gladly followed Karl.

Months passed, and the farm work continued while Berta's tummy grew. Indiana burned in the heat of August. Chores abounded as the boys prepared for the corn harvest. The oat harvest and thrashing had taken up most of July. On this day, Berta and Amelia canned pears for the upcoming winter. Berta sat for a moment, thinking about her canning experience in July. Canning pears seemed like a respite after putting up green beans, peas, and tomatoes, the month before. The green beans were particularly tiring: the bending over to pick them, the fighting off biting bugs, the snapping of the beans, and the sprained backs.

She remembered setting up the canner cooker in the tremendous heat of a July day. That day could make anyone swoon. It was extra difficult for Berta to bear well into her pregnancy. On the twelfth of July, while the men harvested oats, she and Amelia picked tomatoes at night when it was cooler. They soaked the tomatoes overnight, which helped clear them of dirt and bugs. They rose early on the morning of the thirteenth while it was a little cooler and fired up the kitchen stove and began to boil water. On that morning, the outdoor thermometer mounted on the Hoffman's north window read ninety degrees and the indoor thermometer was a degree warmer. The temperature rose as the day wore on, climbing higher and faster in the kitchen. Soon it was ninety-six degrees outdoors, hot and humid enough to scald any Hoosier. The hot stove and boiling water in the kitchen heated the room to a steamy 101 degrees by 10 a.m.

A Winterhalder and Hofmeyr Grandfather clock stood as the guardian of time in the family room of the house next to the kitchen. The sculpted, dark cherry wood held the clock and the long pendulum that rocked back and forth. When Berta heard it chime to indicate it was 1 p.m., the kitchen's temperature had climbed to 106. She fell to the floor with a small, soft thud almost silently. Amelia grabbed a towel, dipped it into a bucket of ice water, and placed it on Berta's forehead. She ripped open the thin, little blouse that Berta wore and used another towel to dab her heated body. Eventually, Berta began to awaken, and Amelia sighed in relief.

She grabbed a pillow from the davenport and gently placed it under Berta's head. "You lay there!" she ordered. Knowing her shouts wouldn't be heard over the noise of the tractor and thresher, Amelia ran outdoors, where the men were hand-feeding oats into the threshing machine with pitchforks. Dust filled the air as the threshing machine flailed the oats to separate the seed from the stem. A flatbed wagon still had about half a load of oats to be thrown into the thresher. Amelia ran to the Hoffman men, screaming over the noise of the machinery. "Berta has fallen," she shouted. "Berta has fallen!" John heard her the second time and dropped his pitchfork, outpacing Amelia as he ran to the house. The boys shut down the thresher and followed.

By the time they reached her, Berta had cooled slightly and sat up with her back against the wall. She covered up her torn blouse and kept the cloth against her forehead, although it didn't seem cool anymore. John ran across the room and put his hands on her shoulders. "Are you alright?" he asked. Amelia took the now warm towel, dipped it in the cool, ice water, and returned it to Berta's forehead. She wet another towel and placed it on

Berta's chest, cooling her and providing a modest cover over her wet, torn blouse. Berta glanced up at John, smiled faintly, and said, "I'll be all right. I've got to get back to the canning."

"No!" both John and Amelia exclaimed. At that moment, the boys ran into the house. They were dusty and dirty, like John. Everyone in the house dripped with sweat. John scooped up Berta, and he moved her to the next room, out of the kitchen and on the shadier, north side of the house. He laid her on the couch, then announced, "I'm going to drive into town and get Doc." Berta futilely protested, but John wouldn't let her, and she knew it.

"You take care of her," John instructed Amelia. He knew it would rain that night, which meant havoc for the oats. He asked Karl and Fritz if they could thresh the last of the oats and move the straw and grain into the barn's lean-to to protect the harvest from the weather. They nodded, knowing that their mother was in good hands with Amelia. Berta smiled timidly and waved the boys away.

John walked briskly out to the car and drove away.

He entered the doctor's office at the back of Doc Bane's home. The doctor could immediately smell the moonshine on John's breath. "Doc," John said in a panic, "Berta has fallen in this damned and blasted heat and canning process. I came to get you to look at her."

Doc, short and chubby with a fading hairline, frowned and said, "I'll take myself and save you the trip back in." He hesitated to confront John directly about drinking and driving. Doc followed John in a horse and buggy to the farm, about fifteen miles away.

When they arrived, Berta was still on the couch because Amelia wouldn't let her move. Doc knelt next to her, opened his medical bag, removed a thermometer, and gently stuck it in Berta's mouth. He raised her arm and checked her pulse, then felt her forehead. He laid his hand on her stomach and felt the baby's pulse underneath. He then pulled out a stethoscope so he could listen to Berta's heartbeat as well as the baby's. After a few minutes, he stood and said to Berta, "You'll be just fine, but no more canning until this heatwave breaks." He reached into his medical bag and handed her a small bottle of aspirin. Take two of these now and two in the morning. He smiled at Berta and said, "These new aspirins are a miracle drug."

 After Karl and Fritz finished the threshing, Karl engaged the well pump to the windmill and took an outdoor shower, with his clothes on and all. He didn't want to offend Amelia with his smell after a long, hot day of work. Engaging the pump to the windmill was a tricky maneuver. In this part of Indiana, the wind blew almost continuously upon the windmill blades, so the pump had to be engaged when the Hoffmans wanted water and disengaged when they didn't. Deft moves were required to complete the process, and a few attempts were often required, especially if Mother Nature provided a highly gusting wind. One often wound up with a pinched finger. Karl and Fritz had mastered this effort at a young age while pumping water for the cattle. It was tricky agricultural mechanisms, which reduced the drudgery of some personal labor, but were not safe in the early 1900's.

Once Karl engaged the pump, he splashed the water upon himself. He huffed as the fifty-degree water tensed his body, a chilly contrast to the day's heat.

That evening, when it was a bit cooler, Karl and Amelia finished the day's canning. Rain plopped on the roof of the house as they finished up around 10 p.m. The rain helped cool the home, and Berta began to feel better. The next day, the temperatures cooled into the 80s, still hot but a welcome respite from the heat of the last few days.

In August, weeks before corn harvest, John took Berta to Indianapolis. The city was undergoing a growth spurt as rural folks moved into town. An Indianapolis banker helped John close a sale on land just outside the city, and John was sure the urban growth would return a good investment.

After the deal was done, they went to a tavern known for good Irish food and whiskey. As soon as they were seated, a young waitress approached them and handed them menus.

John looked at Berta and said, "Let's each order a drink and celebrate today's deal."

Berta's face lit up in embarrassment as the waitress stood over them, and she said, "John, you know I don't drink."

"Okay, be a spoiler," he replied. He looked at the waitress and said, "I'll have a whiskey." Then he turned to Berta. "What do you want, dear?"

"I'd like a cold Coca-Cola with ice, please."

When the waitress left, Berta extended her arms and hands across the table, and John reached out to hold her hands.

"Oh, John," she said meekly. "You know I love you with all my heart, but I need to say something."

John wasn't stupid; he knew what was coming. "Go ahead," he said.

"It's just the drinking. It's not doing you any good."

John reared his head back and answered, "What do you mean? I'm taking care of you, the boys, and the farm. My drinking hasn't hurt anyone."

"It's hurting me," she said. "I watch you slink off to the barn...."

"Do you want me to drink in the house?"

"That's not the point. You know there's a lot of news about how bad alcohol is. Now, you're causing the boys to drink."

"They enjoy it!"

Berta frowned and said, "Maybe that's my point. They are enjoying it and drinking responsibly. You're drowning yourself."

Both became quiet. Berta had said her piece and knew she couldn't say more. John didn't hate her for it because she cared for him. He thought but wouldn't say out loud. 'It's how I escape what I've done.'

The rest of the dinner and the ride home were quiet.

Chapter 4
1920 - 1922

Berta gave birth to Patrick on January 10, 1920. John had subtly suggested that the family start using English or Irish first names. They had originally chosen German names because of the Hoffman family's heritage. After fighting the Germans, John no longer cared for that idea, and Berta agreed.

A week after Patrick's birth, the national prohibition went into effect. Even though Indiana had supposedly been dry since April 2, 1918, no one enforced the law. Moonshine Moore couldn't have been happier. His business skyrocketed. To keep his illegal system quiet, Moonshine donated some of his liquor to the local sheriff and county judge. His increased business allowed him to charge more, which paid for the freebies to that "feedbag for a judge," as Moonshine called him.

Moonshine also sold his hooch to customers in Lafayette. Bar delivered the moonshine, and he was easily recognized with his tall frame, sturdy body, angular face, and white, straw cowboy-style hat. He was good-natured and good at his job. Bar distributed moonshine to several clients in Lafayette late one night shortly after the national prohibition. Men gathered and chatted while they purchased their share of moonshine in an alley behind a general store.

On this evening, a car pulled down the alley with its headlights glaring. That was a no-no for anyone on Bar's list, so the young man immediately knew something was wrong. The car stopped several feet from Bar and his gang of customers. The engine idled for a moment, and then two men stepped out, each holding a Tommy gun in the air. They approached the crowd, a lone overhead light illuminating a few of the men. One man approached Bar and stated, "We're going to take your moonshine, and you're going to stop selling it in Lafayette." Bar cracked a smile and gently shook his head but said nothing. The man continued, "Lafayette is now Al Capone's territory. You won't be allowed to sell here any more."

Bar's smile faded. He didn't know much about Al Capone, but what he did know made him figure these were Chicago thugs. Slowly, he drawled, "And you two think you're going enforce that when you aren't even from here?"

The second man nodded, and then mayhem broke out. There must have been a dozen of Bar's customers in the alley. Four emerged from the shadows behind the two strangers and ripped the Tommy guns from the men's hands. Others quickly wrestled the two to the ground and beat them soundly. The violence ended quickly. Bar stood over them and announced, "Well, you tell that 'Capoone' guy, whoever he is, that this town is NOT his town to control. If you come back to Lafayette, you will die." He grabbed one of the men by the arm, pulled him up, and guided him back to the still-running car. At that, the two climbed in as Bar and the others stood watch with a bevy of handguns and the Tommy guns in hand. Bar followed as the car backed out of the alley. He emptied the chambers of both Tommy guns, opened the back door of the car, and tossed the empty guns inside.

"One thing I am not is a thief," he said, shutting the door. "Don't come back." And with that, Bar watched Capone's men drive away. Good fortune was on Bar's side. The men had acted without Capone's permission, so they never reported the incident to him. Moonshine and Bar's business grew very profitable in Lafayette.

John kept a steady supply of moonshine for himself and his sons. Throughout the winter of 1921, he made regular visits to The Station to play euchre and drink. The cards flew fast; an outsider couldn't keep up. Money exchanged hands at a nickel per hand with side bets on who held what card. Most of the men chewed tobacco and spit into colorful empty cans with names like Sweet Violet, Lucky Strike, and Granger. Rumors that Highway 321, a route between Chicago and Indianapolis, was a thoroughfare for Chicago gangsters were repeated and embellished a little more each time.

At first, when John returned from these outings, Berta would yell at him.

"You are not only poisoning yourself, but you're dragging Karl and Fritz into this mess," she'd shout. Sometimes she strained her throat and would massage it with her long, slender fingers. Most of the time, John would stagger into bed, his snores reaching Berta in the living room. He carried on in the downward spiral that began when he returned home from the war.

In March, Berta put her foot down. Though divorce was incredibly rare, she felt it was her only option. One morning after the boys and Amelia left the breakfast table. Berta told John to remain in his seat. She knew that breakfast was the best time to reach him when he was logical. That side faded as John tipped the bottle, and the day wore on. His wide-eyed expression told Berta that he knew he was in trouble. He said nothing.

"I've been talking to Pastor Duncan," she said. It bothered John that Berta felt she had to talk about their relationship with an outsider. "He understands all that you have been, though. He and I want you to pull through this, so we talked." Berta sat silently for a moment. John was locked up within himself.

Berta continued, "I want you to stop drinking, or I want a divorce." Those words burnt her throat like a fire on gasoline. They both sat silently. The grandfather clock in the next room ticked off the seconds. John and Berta still loved each other. Berta knew that, but right now, she wondered if he loved the moonshine more than her.

When John finally broke his silence, his words were strained. "Well, I guess I could live in the empty rental house and still farm," he strained out the words. The rental house sat about two miles away from Hoffman's home on another part of the cattle farm. Then he stood and walked outside to the barn. Berta burst into tears, sobbing so hard she couldn't see.

A tough woman, Berta had gained her composure when Amelia reentered the kitchen. Berta told her what had taken place.

Silently, Amelia sat next to Berta and held her hands. Divorces weren't unheard of in the '20s, but they came with a heavy stigma. There would be gossip, and people would jump to conclusions and spread rumors. John knew this would hurt Berta. Alone with his thoughts in the barn, John knew he would not contest the divorce. In fact, he took the blame so easily upon himself that Berta began to think he didn't love her anymore. Berta couldn't see that his actions were only intended to save her from more heartache.

John spoke to the boys that afternoon. Before he could tell them that they could stay with Berta or join him, Karl blurted out, "I'm going with you, Dad." Fritz then proceeded into the house to blame his mother for her bold action. She was sobbing again. Fritz froze for a moment, then returned to the outside. Without exchanging a word, Berta knew that Fritz blamed her. He had missed John so much during the war that he quickly became fully engaged in everything that was John.

The second house, or the rental, as John called it, quickly became John and the older boys' home. It was fully furnished for tenants, ready and waiting for them. Here, away from Berta's watchful eyes, they drank moonshine inside the house whenever they wanted. They told stories, chewed tobacco, and spit into empty coffee cans. Fritz, in his last year of high school, often went to school hung over. Sometimes he was still drunk from the night before.

One night, both Karl and Fritz rode to The Station with John. They were drunk. The Hoffmans watched the men who had already gathered, Moonshine and Bar, among them, play euchre. After a few moments, Moonshine slammed his cards down and exclaimed, "Brian Alan Ryan, if I didn't know better, I'd say you are cheating."

Karl smirked. He rarely heard Bar's full name. Karl mimicked Moonshine: "Brian Alan Ryan." The others laughed at Karl's teasing. He continued, mouthing each letter, "the B-A-R of Boulton." One of the men cautioned Karl to be careful. "He's ousted Chicago thugs out of Lafayette."

Bar, good-natured as always, smiled and said, "It's okay. I kind of like it."

After that, Bar became known in the neighborhood as the Bar of Boulton. That nickname even made its way into Lafayette, where some people asked for Bar Moonshine instead of Boulton Moonshine. That pleased Moonshine Moore and Bar because it kept their location hidden if the federal agents came looking.

Of course, John and his sons worked on the farm every day. Often, they visited with Berta and Amelia. Karl cleaned up his act when he was in the presence of Amelia. Over time, the women saw that none of the men were eating properly, so they offered breakfasts and lunches when the men came to the farm.

If not for the coldness of their interaction, one wouldn't realize that John and Berta were divorced. The two kept their comments short and to the point, limiting their discussions to farm work. If anything needed fixing in the house, Berta asked Karl to do it. It was during this time that Berta noted a special interest growing between Karl and Amelia.

One evening, after the men returned to the rental house and Patrick was in bed asleep, Berta broached the subject. "You and Karl are hiding your relationship less and less every day."

Amelia smiled and asked, "Is it that obvious?"
"To me, it is. How long has this been budding?"

Amelia rubbed the back of her neck. "I think it really started when Karl, Fritz, and I went to the movie in Lafayette. It was an anti-war movie, but more importantly, it was a love story. It was called "The Four Horsemen of the Apocalypse." Amelia hesitated, then continued, "There was an actor, Rudolph Valentino." She glanced at Berta and added, "He was handsome, but throughout the whole movie, I only thought about Karl. How I enjoyed going to this movie with him, but also about how he makes me feel every day."

"Have you two talked about this?" Berta asked politely. She led Amelia to the kitchen table, and the two of them sat.

"No, not really a word, but I'm beginning to think about telling him how I feel."

Berta hesitated. "Now, I don't mean anything against you, but have you considered that he is two years younger than you? He's just a boy."

Amelia laughed hard. She quickly regained her composure and said, "And I don't mean anything against you, Berta. But you are his mother. He will always be your boy. But he's nothing but a man. Have you seen how others respect his manliness? I have."

At that point, Berta decided to dive into her genuine concern, "Yes, I am his mother, and I only want his happiness. Your happiness counts too, but have you considered how much he is like his father?"

Amelia longed to think that John's drinking problem would pass and that John would return to who he truly was. Karl's drinking wasn't as extreme as his father's, and Amelia thought he never would be the same as John.

Berta continued, "I can't promise you that we all will get over this phase."

"Oh, we have to," Amelia replied.

Berta turned and looked at the kitchen door and mumbled, "I just don't think we will move on from this."

Chapter 5

October 1922

John sat with Karl and Fritz on a hot evening that October. They participated in their favorite pastime: watching the sun set while drinking moonshine.

As the sun moved lower toward the western horizon, the great white orb seemed to grow and turn a brilliant burnt orange. The scattered clouds reflected bright flares of sun dogs that flashed out like spokes in all directions; the blue of the surrounding sky was afire with hues of purple and pink.

While looking at the sun, John mused. "I've been wrong calling you boys." They all laughed, but then he continued. "Well, I mean, you'll always be my boys, but I have to admit you've become men. If I ever refer to you as boys again, hit me upside the head with a two-by-four."

They laughed again.

After a few moments, Karl said, "Have you ever noticed how some people know what others are thinking?"

"No one can know what another person is thinking," John replied.

"Like telepathy?" Fritz asked.

Karl frowned. "That sounds something like one of your new words. I'm talking about when someone is so in tune with a person, they know what that person is thinking."

John shrugged his shoulders and said, "Hmm, I never thought about that."

Karl continued, "Mom is like that. She reads me so well; she often knows what I'm going to say."

John and Fritz slowly nodded. Fritz said, "Yeah, I get it. She does the same thing to me."

John took another sip of moonshine and said, "I guess I'll have to pay more attention to what I'm thinking in the future."

They all laughed again.

The men had settled into a routine. For their suppers, they usually ate beef and vegetables from the farm. In the spring, summer, and fall, the three worked with Berta and Amelia in the garden and orchard to put up vegetables and fruit for the brutal Indiana winters. Berta and Amelia continued to prepare breakfast to help them all start their workday. Berta never fixed dinner because the men always retreated to the rental house for a night of drinking. While Karl and Fritz fit into the routine like breathing, John was miserable. The more misery, the more he drank. Sometimes he could barely function at breakfast. It hurt Berta to watch his demise. She blamed it entirely on the war. John was not the same man she married and who fathered her children before the war.

John was either drunk or hungover that entire harvest. The Hoffmans used a corn picker pulled behind two Percheron horses. While this was better than harvesting by hand, corn pickers clogged easily. One day, John was thoroughly drunk while leading the horses. Karl, who was rotating the wagons into which the picker emptied corn, noticed that John was not handling the horses well, and the picker clogged even more than usual. He watched his father stop the horses and go back to unclog the picker. John stumbled and fell toward the gathering chains. Karl immediately ran to his father's aid and saw John's arm become pinched by a roller chain, blood spewing. John gathered himself and stood in front of the picker, squeezing his arm tightly. Karl ripped off John's sleeve and wrapped it around the wound, and then placed his father into the empty wagon and took him to Berta's house.

Berta saw John lying in the back of the wagon and ran to see what had happened. John was so drunk that he had passed out. Berta helped Karl move John to the Model T.

"You go back and help Fritz with the harvest, and I'll run him into the doctor's office," Berta told Karl.

John remained unconscious for the entire trip, only awakening when the doctor roused him. Together, Doc and Berta guided John into the office.

"Looks like you've cut yourself pretty badly, " the doctor told John.

"This little scratch?" John replied. "I've seen a lot worse."

Doc Bane nodded, certain the war veteran had indeed seen worse. He placed fifteen stitches in John's wound. When he finished, he looked John in the eyes and said, "You've got to start taking care of yourself." Berta shook her head in disappointment and led John back out to the vehicle and returned him to the rental house.

As difficult as this new lifestyle was for John, it was doubly worse for Berta. When John had left for the war in 1917, she never dreamed that she would face years of lonely nights in bed. That nighttime loneliness was harder on Berta than John because John usually passed out by the time he went to bed. The loneliness was only one difficulty Berta faced. She was tough and spirited, and she needed to be every time she made the fifteen-mile trek into Boulton for needed supplies. In town, she was known as *the divorcee*.

Some women in town pointed up their chins and turned their heads when they crossed Berta's path. Others could be heard mumbling, "She should be ashamed of what she has done to that Hoffman family." Amelia, who always traveled to Boulton with Berta, would gently put a comforting arm under Berta's and guide her away from the judgmental people. Over time, Berta avoided travel into Boulton in the afternoons when some of the men had been drinking. She couldn't forget one man's surprised reaction as she walked down the wooden sidewalk. As he emerged from an establishment that sold Moonshine's hooch, the man shouted, "Do you still go by the name of Hoffman, you slut?" Then he stuck out his tongue, grabbed his crotch, and shouted, "How about some of this? I heer'd you like other men."

Amelia immediately grabbed Berta by the hand and led her across the street to the sheriff's office to report the man who had accosted Berta. Berta protested, but Amelia would not relent. Berta knew Sheriff Bucklyn was in everyone's back pocket, including Moonshine Moore's, so she protested that they should leave. Eventually, Bucklyn invited them into his inner office, his eyes widening as the beautiful redhead and her companion entered the room.

"What can I help you with, ladies?" Bucklyn asked.

His cordial greeting put Berta at ease. Still, she stumbled and stuttered. The incident had shaken her core like the quivering of the ground in an earthquake. "I..w...we are here to report..." she hesitated. Bucklyn opened his mouth and cocked his head as if to say, 'Well, get on with it.' Berta continued, "...an incident."

Bucklyn leaned forward with his elbows on his desk and clasped his hands. "What is the nature of this incident?"

Berta glanced at Amelia, then back to the sheriff. "Well, sir, I have been assaulted. A man came out of that place over there and shouted at me." Berta pointed out the window and across the street.

Bucklyn looked at his watch. He was pretty sure who would be drunk at this time of day. "Was he short, round-faced, with red cheeks and curly hair?"

Berta glanced at Bucklyn; Amelia grabbed her right hand. Berta nodded a timid yes. She wondered why she felt so uncomfortable, so intimidated. At home on the farm, she would have stood up for her rights. She would have grabbed the shotgun.

Berta gasped with surprise when Bucklyn laughed. "Well, hell! Excuse my French, but that's Bud. He's as harmless as a fly on a cow's ass. Just ignore him."

Taken back, Berta retorted, "Well, he made a sexual gesture toward me." The words made her even more uncomfortable.

Bucklyn paused. What came next stunned her. "Well, Mrs. Hoffman." He paused, then continued, "I can still call you that, right? I'm trying to be kind here, but you must understand that a woman of your nature brings forth behavior in men like Bud. You are as responsible as he is for what took place."

Berta closed her mouth and glanced at Amelia. Regaining her usual composure, Berta said, "Let's get out of this...this...." She wanted to say pigsty, but she refrained. "Let's get out of this town," she finished.

Berta abandoned the rest of her shopping. Hurt and distracted, she could no longer focus on her list. Amelia followed her to the Model T pickup. They couldn't get out of Boulton quick enough and were silent the rest of the ride home. Berta decided not to tell John about the incident.

Two weeks later, John drove into Boulton to pick up Frank Trevelyan. The two had corresponded regularly after the war. Today, Frank stood at the train station waiting to meet his war buddy. Returning to the farm, John drove west into the setting sun. Frank pointed across the wide-open sky streaked with beams of red, blue, white, and pink. "Now that's something I don't see much of back home," he said. "Too much pollution from London. Sometimes the burning coal blackens our sky." He turned and looked at John behind the driver's wheel. Then he returned his gaze to the sunset and said, "Now that's a God-given blessing."

That night, Karl, Fritz, and John sat up late with their visitor from England. He stayed with them in the rental, which meant he would sleep on the couch. Karl broke out the moonshine.

Frank had never tasted moonshine, but he took to it rather quickly after it burned his throat with his first sip. "That would put hair on a young man's chest," Frank barked.

"Maybe that's what happened to Fritz!" Karl exclaimed. Frank, a naturally jovial man, burst out laughing.

John noticed that Frank had gained a lot of weight since the war. "Looks like the post-war life has been good for you."

Frank patted his stomach, "Thanks for noticing." He looked at his now empty glass of moonshine and asked, "How do they make this?"

John quipped, "No moonshiner is going to tell you his secret recipe." Then he added, "I'm guessing that it is 100 percent of a special food-quality corn, not the corn we feed our cattle. Well, anyway, a corn whiskey recipe is how moonshine is supposed to taste."

Frank looked at Karl, who poured more moonshine into Frank's glass. He considered the liquid. "This is smooth. It has a rich aroma and is powerfully flavorful. Now that you mention it, the corn comes through loud and clear."

John said, "That's what makes this stuff dangerous because it tastes less powerful than it actually is."

Frank looked at Karl and Fritz and asked, "Did your dad ever tell you what he did in Bailleul?" he almost shouted.

"Frank!" John protested.

Frank wouldn't have any of John's protestation. "Then I'll tell you," he said. "Maybe letting the story out will ease that horrible tension wrapped around your papa's neck."

Frank then related the entire story of their efforts in Bailleul. John felt a stirring, a change. He couldn't tell his family, but Frank could. Everything he'd kept bottled up inside was spilling out through Frank, of all people. The blunt Englishman simply speaking his mind was helping John unravel all the misery that he had been hiding.

As Frank described the danger they had been faced, Karl and Fritz occasionally looked at their dad, eyes wide, as if to say, 'Why haven't you told us this story?' Frank described the garden as they had seen it under the German flares. He described Barry, who took the fatal shot to the head.

"After the war was over, many of us drank and partied in Paris. Not your dad. He went back to Bailleul to see if the garden was still there," Frank said.

"Was it?" Karl clung to Frank's words like stink on cow manure.

"I'm afraid your dad's coordinates were too wide an area. Not only did he send about forty Jerrys to hell, but he also wiped out the Jardin Botanique that his ancestors had loved."

Moments passed as the men sat quietly.

Finally, John broke the silence. "That Jardin was a couple of centuries old, and... and I, I had it blasted off the planet. I can never forgive myself. My great uncle must be rolling in his grave."

Frank gave John a stern look. "The Jardin? The Jardin? After all, we've been through, that's why you are torturing yourself? Hell, you've got land, money, and green thumbs. Rebuild it yourself."

Suddenly, John froze in place. Frank's simple suggestion had never crossed his mind. That suggestion changed John like a desert sojourn might change a world traveler. He looked at his drink as if it were a knife at his throat. He set it down and considered Frank's words.

While John sat in silence, Frank felt the moonshine pressing hard on his bladder, and he went outside to relieve himself as he'd seen the Hoffman men do. Stepping away from the house and into total darkness, Frank stopped in his tracks. He was overwhelmed by the sight of the Milky Way Galaxy and the planets overhead. Still, he administered to the task that brought him outside but continued to gaze up at the sky. He thought about the sunset and the view above him and thought: They are truly blessed to live here.

The next morning, thick with a hangover on his brain, John pressed on in a field near the farmhouse where Berta, Patrick, and Amelia still lived. He stepped off the space along the home's driveway. John thought, "That's two hundred paces and more than enough room." This field was lush with pasture; it would be perfect to convert. Karl, Fritz, and Frank drove the converted Model T pickup up the lane. Frank asked Karl to stop where John stood. He climbed out and shouted, "Now what are you thinking about?" even though he had an idea. Starving, Karl and Fritz drove to the farmhouse to eat breakfast and begin their morning chores.

John was lost in thought; his right arm was extended with a thumb sticking straight up. He imagined each satyr at the corners of the field.

Frank interrupted his thoughts. "Say, ol' boy, what are you up to on this humid, hot October morning? It never gets this blasted hot and humid this time of year back 'ome."

John nodded but didn't speak. After a moment, he simply stated, "I can rebuild it here."

Frank chuckled. "I'm happy that you like my idea, but surely you're still drunk."

"No, no," John said. "I can see it in my mind." He hesitated, then added, "If I can see it, I can rebuild it here. There's more than enough space, and the soil's a perfect match."

Frank chose his next words carefully because he didn't want to offend John. "Well, you have a green thumb," he said, looking around the lush cattle farm. But the statues, how will you rebuild the statues? The expense?"

John nodded, holding one index finger up in the air to quiet his friend. "That will be tricky, but the expense will be no problem. The Hoffman's estate has plenty of assets, and with the expanded trade into war-torn Europe, the farm is very profitable right now." John didn't mention the investment property in Indianapolis. He stood quietly and imagined the dream that this field would become. The decision was made. He patted Frank on the shoulder and said, "Let's go up to the house and get some coffee and breakfast."

Frank snorted, "Well, at least what you Americans call coffee."

John ignored his friend's teasing. He sensed that a new day was dawning, not only for the new Jardin Botanique but for John's healing from the war.

Frank stayed for the rest of the week. The next morning, John asked Karl to call the investment banker in Indianapolis. Standing nearby, John waited pensively. When Karl handed the phone to John, the elder Hoffman accidentally turned the phone handset upside down and began to talk into the speaker end. Karl turned it around for his dad. Frank shrugged and joked, "That explains why he never calls me."

John had trouble hearing the man at the other end of the line. "William?" John almost shouted into the phone. "You know that contractor you mentioned a while back who is building the war memorial in downtown Indy?" No one could hear the response, but John motioned for a pencil and paper. "Wait, repeat that," John said into the telephone and scratched out a name and address. "Thank you, William," he said and finished the call.

John wiped his hands on his shirt, looked at Frank, and said, "Let's go to Indianapolis." On the ride down, John explained that the city was investing in fountains and memorials to honor those who had served in all of the wars, from the American Revolution to the Civil War and now the Great War, in which he had fought.

"I've seen his work, Frank. I think he will be perfect for building the nymph and the satyrs." John could barely contain his excitement.

"Well, I'm glad that I could inspire you," Frank said. Then the vehicle hit a rut in the road, and Frank bounced on his seat. He noticed that John was oblivious.

Chapter 6

October 1922

At the end of October, John looked over the leveled field. He had dreaded plowing up the pasture, but he knew the result would be spectacular. For three weeks, twelve hours a day, he leveled the soil by pulling the straight trunk of a heavy cottonwood tree behind two Percheron horses. Winter would soon be upon them, so he made passes over the land day after day to achieve a perfectly flat grade, with Karl and Fritz lending a hand whenever they could. When the field and soil were ready, John planted winter wheat. He had no intention of harvesting the wheat; it would be used as a winter cover until he could plant grasses and flowers in the spring.

John made many trips in the Model T to a Lafayette nursery to purchase trees. He remembered the French Jardin displayed six cherry trees, which were in place by November. He labored little with his memory of the Jardin even though more than a decade had passed since he had seen it in all its beauty. Berta, and often Amelia, watched from the turret of the house. When John refused to stop for water or lunch, Berta brought refreshments to him. Sitting on the back of the pickup wagon, John described his project. Berta understood what he was recreating but could not envision what John held in his mind. She decided just to watch and wait for the result.

Berta noticed the change in John; she sensed that he was drinking less – or not at all – since Frank had visited. She thought John had returned to the man he was before the war. As she watched him quickly snarf down his lunch, Berta thought about how passionately the John of old had lived. She smiled inwardly, thinking that he was no longer passionate about liquor.

Karl began to spend more time at the farmhouse with Amelia. She had been born in the Kingdom of Bavaria, now a free state after the Great War. Amelia was the eighth child born to a family during a time of strife. Her mother had died in 1910, leaving her daughters to shoulder the household responsibilities. Amelia, the youngest, dutifully attended to her chores but knew she would never be more than a shopkeeper's assistant. She dreamed of the new world and the promises it held.

In 1915, at the tender age of fifteen, Amelia departed for America and met relatives in Indianapolis who had immigrated from Germany. She hoped to work and save enough money to own a business and live the American dream. Though plenty of work was available, factories brutalized women into drudgery and low pay. After two years of harsh factory work, her dream stayed well out of reach. As luck would have it, a coworker knew that her cousin Berta Hoffman wanted a house maiden and was certain that Amelia would be a perfect fit. She also knew that Amelia would be in good hands with the Hoffman family. Some day she might be able to start her own stitchery shop in Boulton.

Boldly, at the suggestion of her co-worker, Amelia quit her job and headed to Boulton for a job interview and, hopefully, the next step in her journey. Berta met the young woman at the train station then offered to treat her to a cup of coffee across the street. Amelia, familiar with American coffee, timidly asked Berta if she could have a chai. Berta smiled, patted her hand, and said, "Of course, you can." When the waiter approached their table, Berta said, "We will be delighted to have two chai, please."

The two women talked for more than an hour. Eventually, Berta asked, "How old are you?"

Amelia, still timid, barely said, "I'm seventeen. Is that okay?"

Berta thought about the proximity in age to Karl, who was two years younger, and she considered Amelia's petit and beautiful presence. She wondered if that would be good or bad, having a woman two years older than her son in the house. Eventually, she laughed and said, "You seem so much more mature than my son Karl. But..." She hesitated and then said, "...but you have so much worldly experience over him." Amelia presented herself as a young woman in her mid-twenties, and Berta realized she enjoyed the thought of a younger woman in the Hoffman home.

Finally, Amelia exclaimed, "Must I take the five o'clock train back to Indianapolis?"

Berta looked up with surprise. "I'm sorry, my dear! You've had this job based on my cousin's recommendation alone!" Again, she smiled. "The real question is will you have us? If so, grab your bags, and let's take you to your new home and see about the answer to that question."

By December, winter had blasted its way into northwest Indiana. John moved from grading and planting to planning for the Jardin.

It turned out that Amelia drew very well and provided an artistic representation of the statues. Both John and Amelia turned red-faced when John described the satyrs' penises. Berta snorted, "Grow up, you two. This is a cattle farm, after all. Nothing you haven't seen, Amelia."

John traveled to Indianapolis and showed the drawings to the contractor. "These are fantastic!" he said. Then he hesitated for a moment and looked at John. He continued, "I don't know you from a hill of beans. But I must warn you that this will be very expensive."

Without hesitation, John plopped $1,000, a fortune in those days, onto the construction table. He had prepared for this moment. The costs of the public statues and fountains being erected in Indianapolis were a matter of public record. John had done his research, and he estimated that this would cost about $2,000. John simply said, "How about half now and half when finished?"

Surprised but grateful, the contractor told John, "Please have a seat. I will prepare a contract for our business and a receipt for your cash."

John's work on the Jardin hadn't just attracted the attention of Berta and Amelia. Others, primarily locals out on Sunday drives in either horse and buggies or cars, came to investigate the drastic changes and speculate on John's motivations. Green still dominated the field as the winter wheat grew, but John kept the rest of his changes under wraps, adding to the mystery.

Returning from the contractor's office, John stopped by the farm to check on Karl and Fritz. Berta told him that they had retreated to the rental home for the evening, then offered John some warm coffee to fight off the cold of his trip. The silent house seemed strange to John. "Where's Amelia and Patrick?" he asked.

"Oh, Patrick has been asleep for about a half an hour, and Amelia took the pickup over to the rental with Karl and Fritz." She smiled and continued, "She and Karl can't seem to get enough time together." John thought quietly for a moment. It seemed strange that a woman would go to another man's house. Berta laughed and said, "John, I know what you are thinking. They've become quite attracted to each other, right under your nose." She hesitated. "I'm sure they will be fine."

They both smiled. The family had become very fond of Amelia, especially Karl. Berta touched John's arm with her long fingers, stroking his flannel shirt. "We have a problem," she said. "There's a raccoon under the house. It's been making a terrible racket for the past two nights. I think it's making a nest for the winter."

It wasn't unusual for homes in the country to have raccoons underneath. John nodded. He knew just what to do. He took the last sip of his coffee and then grabbed his coat hanging next to the door. "I'll be back," he said, retreating to the barn. He hesitated for a moment and looked at the shelf where he had kept the moonshine. He admitted the desire to drink it still burned in him, but he rejected it. He climbed upstairs to where they kept the live traps, carefully carrying a lantern to light the way. Soon he held the trap by its handle and walked back to the house.

Setting it down outside by the front door, he walked in to find Berta standing by the door with her hand extended. In it, she held a squished patty of sweet corn and fish from the evening's supper with Karl, Fritz, Patrick, and Amelia. "So, you already knew that I would set this trap," he said with a chuckle. Berta gave a sideways smile and nodded. John retreated outdoors again and placed the trap under the house.

In the meantime, Berta used the few minutes that he was outside to freshen up for John's return. She applied a small amount of perfume and donned a lacey blouse that, while beautiful, chilled her slightly, a sacrifice she willingly made.

When John returned to the kitchen, Berta sat across from a warm slice of apple pie. The aroma of the pie and the whiff of perfume enticed him. Hungry, he sat next to Berta and pulled the piece of pie closer. The tastes and scents enveloped John in a comforting sense that he was back home with his Berta. He thirstily drank the glass of lemonade she had poured.

Again, Berta stroked John's flannel shirt with her long fingers. She started to ask if he would stay the night, but John knelt beside her chair and put his head on her lap. Passion burned in both of them as she stroked his hair. After a few moments of silence, John raised up and kissed Berta softly on the lips. He massaged her thin but sturdy shoulders. They kissed passionately and repeatedly; their hunger for each other equal. John thought about doing more right there in the kitchen. Finally, he asked, "What about Amelia? When will she return?"

"Oh, John, someday you'll learn that women cooped up in the same house all day have our way of communicating. By now, she's talked Karl into a place to stay for the night." John looked at Berta, who said, "Not the way your dirty mind thinks. They are good kids. I'm sure Karl will let her sleep in his bed while he sleeps on the couch. Trust them." John picked Berta up and cradled her long body in his arms. She gently put her arm over his shoulders, and they retreated to the master bedroom.

At the rental house, Karl, Fritz, and Amelia talked by the flickering light of candles and kerosene lamps, their shadows bouncing on the walls and ceiling. Their conversation similarly danced from one topic to the next: John's project, John and Berta's relationship, the farm. Karl and Fritz drank moonshine but refrained from chewing tobacco out of respect for their guest. Fritz, curious about Amelia's travels and life overseas, then asked her about the latest word of the day he'd read in the Chicago City News. "Have you heard the word ritzy? Do you know what that means?"

To the surprise of both brothers, Amelia replied, "When I traveled through New York City, I saw this beautiful, new hotel called the Ritz-Carlton. I was told all the fancy people hung out there. So, does it have something to do with that?"

Fritz replied excitedly, "You are exactly right! This new word, maybe a slang word, is derived from *that* hotel. It means elegant!" As Amelia beamed with delight, Fritz continued, "I've read about the Ritz. Did you get to go inside? I hear it is splendid."

Meanwhile, Karl remained stoic. It sometimes bothered him that Fritz and Amelia knew so much about the world outside of Boulton, Indiana. This was one of those times. However, he knew the feeling would pass when he returned outdoors to continue his chores tomorrow. In that, he was unequaled.

Amelia read Karl like a book. He wasn't worldly, but she adored his work ethic, one that she possessed herself, along with his love for his family. The fact that no one paralleled him in his physical looks and strength helped, too. Not even Moonshine's helper, Bar, who was strong and tall but not as handsome as Karl, had caught Amelia's attention. She attempted to deflect the grandiose attention from herself. Looking back at Fritz, she exclaimed, "Oh no. They wouldn't let us, peasants in the Ritz, but we could see how grand it was from the outside and by the way people who went inside dressed."

Fritz continued, "Germany became a world powerhouse after unification, before that war, that is. Only forty percent of Germans lived in rural Germany by 1910, a drop from sixty-seven percent at the birth of the empire." Karl glanced at Fritz as if to say, 'How do you know this?' Amelia, however, looked pleased that Fritz knew something about her home country. He continued, "Industrialization was the key. How modern! What was it like to live there?"

Karl interrupted, "They didn't get everything correct!" He raised his eyebrows, insinuating the outcome of the war.

Amelia insisted that Karl was correct. "Too many people tread on lower-level families like ours. The aristocracy didn't understand the lower class or the common rural needs. It was all about urban growth, which lined the pockets of the rich." She was quiet for a moment. "Then there was my mother. For some reason, Bavaria experienced many travelers from China; I think it had to do with the spice trade, at least in Mother's case. Mother helped my dad run a shop. It featured many luxurious food items and spices. After all, in 1910, Germany benefited from economically fantastic times." She pressed her hands together in her lap. "That wouldn't help my mom. She encountered Chinese visitors carrying the pneumonic plague. She contracted it and died before 1910 was over."

The men were silent. Amelia wiped a tear from her eye with the back of her hand. Karl chided Fritz, "Now see what you have done!"

"It's all right, Fritz," she said. "That story is part of who I am. I can't go back in time and change it. It's also part of why I am here today. This part of the world is not overcrowded like Bavaria had become. And I like the people here."

The three talked for another hour. As it grew late and John remained at the farmhouse, they knew that Amelia would stay there for the night. Karl said, "My bed is clean and made up. You can stay in there tonight." Amelia began to protest. Karl continued, "The couch is no place for you. I'll stay out here." With that, they ended the conversation and extinguished the kerosene lamp. They retired, each with their own candle and thoughts spinning in their heads.

Karl wondered if his mother and father could repair their marriage. More pressing, his mind turned to the thoughts of sitting up and talking with Amelia late into the night. He liked that. Fritz moped as he went to bed and apologized to Amelia for bringing up hard times. She forgave him again. Amelia felt as if she had stepped into the future in Karl's bedroom. She slipped into Karl's bed and smelled the pleasant masculine scent of the man who usually laid there. She wrapped herself in the sheets and determined that she and Karl would be husband and wife.

Chapter 7

1923

Berta bore Samuel in September of the following year. Never disappointed in another child, she hid her wish that Samuel was a Stephanie. Patrick, now three, took to the new baby immediately. He often held Samuel's head in his lap and sometimes napped with him.

Berta and John chose not to remarry at this time, and of course, this meant that the locals talked about them. Moonshine, disgruntled that John no longer drank and no longer buddied with the men, complained first. When the men played euchre in the backroom at The Station, they all talked about the Hoffman's. Moonshine commented on the fact that while John and Berta had been married with the first three children, Samuel had been conceived and was born out of wedlock.

"It's a shame what that woman has done to that man," said Moonshine during the game. "A decorated soldier, he used to be a hard workin', hard drinkin' man, like us. Now he's going off like some crazy fool building that thing and not 'tendin' to the farm as he should." The other men nodded.

"A git is what they have there," someone else said of Samuel.

Moonshine added, "The church and the state should not recognize that git. Nothing good will become of a child born out of wedlock." The men grumbled.

John and Berta, busy with their new child, stayed away from The Station. For a while, so did Karl and Fritz, who attended to harvest instead of drinking and playing euchre. And for a time, the Hoffmans were oblivious to what others thought.

Just after harvest, Fritz drove to Indianapolis for the last satyr. Although it was cold, he basked in the sunshine streaming through the windshield and the warmth of the newly installed heater. He marveled that the heater drove out the freezing cold he'd always been accustomed to for many winters. The Hoffmans had modified one of the hay wagons to carry the satyrs and parts of the fountain back to the farm. On this day, picking up the last piece, Fritz carried $1,000 in cash from his father, the final payment. The bills weighed on his mind during the one-and-a-half-hour trip to Indy. Matter-of-factly, he thought about the woman of his dreams and what they could do with that money.

For several minutes, Fritz thought of that woman in Indianapolis, Elizabeth Lee. He was smitten the first time he patronized the diner where she worked. Her soft, slightly curved chin fit her oval face and perfectly oval lips. Her eyebrows danced with enthusiasm on eyes that penetrated a man's soul. Known as Lizzy, the young woman was accustomed to male advances and intentions, and therefore she owned an advantage over Fritz, just another Hoosier boy from the farm. Fritz remembered that she dressed as a modern city woman with shorter skirts than he'd seen on the streets of Boulton. Her shimmering nylons burned into his memory.

The first day he visited the diner, she greeted him with a tight, wry smile and a slight turn of her face as if to hold him under consideration and, yet, offer him a sense of wonderment. From that day on, each time he returned to Indianapolis, Fritz returned to Lizzy's diner. Eventually, he began to stay overnight with the woman he wanted for the rest of his life. Before departing, Fritz, the ultimate dresser, usually took time at Lizzy's place to match his appearance with the joy he felt. Berta and John knew about Fritz's overnight visits were with a city woman, but only Berta expressed her displeasure.

A car passing in the opposite direction honked at Fritz as his vehicle neared the center line, the wide wagon infringing on the other car's lane. Fritz quickly corrected the vehicle and continued his drive to Indianapolis. Still, he couldn't help thinking about what he and Lizzy could do with a thousand extra dollars. Fritz already had some money from farm profits. They could move to Chicago and enjoy the city life that he read about in the Chicago City News. The wind gusted against Fritz's vehicle and pulled him out of his reverie again. That evening, he proceeded to Lizzy's place. She had the night off, so he arrived in time to clean up, and the two of them went hoppin' to and from basement parties, where Fritz introduced her friends to Boulton moonshine.

Although his mindset toward beautiful women, such as Lizzy, put Fritz at a disadvantage in relationships, his banter about city life, alcohol, and the ultimate buzz played in his favor. An unexpected outcome of speakeasies, which opened to women more than pre-prohibition saloons, resulted in more women drinking more alcohol in public. Before prohibition, most women, if they drank at all, enjoyed only a small amount of sweet wine. By 1923, adventurous women snuck into speakeasies and imbibed gins, rums, whiskies, and wine. Lizzy epitomized those women. She regaled the others with the new names of cocktails and concoctions such as devil's candy and bathtub gin. Fritz, always by her side, never noticed other men's sidelong glances at Lizzy nor the looks she gave in return. As always, her eye captured men's attention. Still, people noticed the pair, Fritz with his stories and moonshine, and Lizzy, who could tempt the most celibate monk.

On this latest visit to see Lizzy, the pair ended their night around 2 a.m. at the Blind Pig, located in the basement of a café on Washington Street. The owner, Brad Owens, had noticed the pair before. On that night, Owens offered both a job at the Blind Pig. Fritz should be the bartender and storyteller, he said, and Lizzy, a new waitress.

Fritz's eyes grew wide, and his jaw dropped. "Ya' gotta be shittin' me!" he exclaimed.

Brad smiled and said, "You two have already been good for my business just by showing up. Imagine how that will be if you're regulars."

Fritz retorted, "Are you going to remember this offer tomorrow when you're sober?"

"You bet! Now, you'll have to work on tips alone, but I think you will do very well on that."

Fritz looked at Lizzy, who smiled and nodded. Fritz extended his right hand and curved it around to shake Brad's. "It's a deal," he said.

The next day Fritz and Lizzy woke up late. It was nearly 11:30 a.m., and Fritz frantically jumped out of bed. Instead of taking his usual time to polish his look, he threw on his clothes, kissed Lizzy goodbye, and headed for the contractor's office to pick up the last satyr. The days had grown shorter, and this late start meant that he wouldn't arrive home until close to dark. John and Karl wouldn't appreciate this, especially Karl, because it meant he had to do Fritz's afternoon chores.

When Fritz showed up at the contractor's office, he nervously hurried the men to load the last satyr onto the wagon so he could get on down the road. When Fritz walked into the inner office, the contractor noticed the young man's disheveled appearance, and he smelled alcohol on Fritz's breath. Frowning, he said, "As a proud member of the KKK, I'm very much against the use of alcohol. Should I call the police so they can investigate?" Fritz frowned but smiled inwardly because he had sold all the moonshine to the owner of the Blind Pig the night before. Nevertheless, his concern for the KKK kept him from nonchalantly dismissing the contractor.

Along with prohibition, the resurgence of the Ku Klux Klan had caught fire across Indiana. It spread from southern Indiana and blasted its way into Indianapolis. Many joined the KKK because they hated the influx of any new people into Indiana; others were coerced. Some businessmen signed on to avoid being run out of business or even out of town. The KKK sought to keep alcohol out of the Hoosier state, limit black migration, and prevent the Catholic church from overtaking what had largely become Protestant lands. In the early twenties, the Hoffmans tried to ignore this second rising of the KKK. John felt the only concerns they should focus on were personal issues. That was a more challenging undertaking for Fritz, who read the city newspapers and visited Indianapolis whenever he could.

Fritz thought for a moment and then said, "The police won't find anything." He reached into each coat pocket, pulled out John's $1,000, and laid it on the contractor's desk. "How about I make the final payment, and I just get out of here and never come back," he said.

The contractor slowly picked up the money while he considered his options. To call the police and then find no hooch would lessen his credibility with the local officers. He quickly wrote a receipt for the cash and then slowly nodded toward the door. Fritz wasted no time leaving the contractor's property.

The ride home jostled Fritz as he thought about the events of the previous twenty-four hours. Of course, he would not take his father's money, even though he thought the Jardin was a waste. Fritz already had saved more than $1,000 from his share of the farm profits, something he would need in the future. Without regret, he thought about leaving the farm. City life called him, just as it did other farm boys, and he answered the call with glee, even with a splitting headache from the night before. He did, however, regret getting home later than intended and having to inform his parents of his decision.

Fritz considered the heat wave they had endured that summer. It had taken a toll on the forage and grain harvests, and Fritz realized the family would need to sell some cattle in the short run due to a lack of feed. He began to form his argument that taking an off-farm job would have less effect on the family because there would be fewer cattle to feed. He also realized that John would have more time for chores with the completion of the Jardin. His mind occupied with these thoughts, Fritz drove obliviously, failing to notice the shocked expressions of his fellow drivers when they saw the satyr's erect penis.

Instead of pulling up to the Jardin to unload the satyr, Fritz drove straight to the rental house to change clothes. Having made great time on his return trip, he figured he could take a few moments to clean up. Fritz completed his detour to his home and approached the site where the last satyr would stand. Shutting the engine off, Fritz jumped out of the vehicle and feigned a smile at John and Karl. John shook his head and asked, "Did you get lost? We've been waiting on you." Fritz shrugged his shoulder without answering directly.

The three men used levers, pulleys, and an overhead frame that John and Karl had placed where the satyr would be unloaded. Carefully, they tied ropes around the statue and then began the long, slow process to lift it into place. As they pulled the rope that Fritz had tied, it unraveled, and the satyr's head struck the frozen ground. Fritz leaped toward the statue. In that split second, he wondered about the problems that would result from a damaged statue and the financial and emotional costs of returning to the contractor after today's incident. He mentally formed his defense of why Karl should return the damaged satyr and pick it up once repaired.

Luckily, the men found no damage to the statue, but John shook his head in a silent reprimand for the lousy job Fritz had done securing the piece. The men struggled to upright the statue and put it in its place. Just after dark, the job completed, the three men looked upon the statue by the light of two kerosene lamps. Unfortunately, the lamps didn't shed light across the Jardin, so John would have to wait until sunrise to view the completed Jardin in its entirety. He motioned for Fritz to take the pickup and wagon up to the barn. "When you're done, come into the house and warm up with some coffee."

Fritz soon joined the rest of his family and Amelia in the kitchen. Exhausted, he hesitated to raise the topic of his move to the city, but he wouldn't be able to sleep tonight if he did not do it now. As they sat around the kitchen table, sipping coffee and nibbling on Amelia's chocolate cookies, Karl held one up and declared, "This is the best cookie I've ever had!" He suddenly froze and looked at Berta and added, "Sorry."

She laughed and replied, "No need to apologize." She took a small bite of her cookie and said, "Amelia is the best!" Amelia turned red, and Berta thought about how close she and Karl had become. Recently they had begun sitting beside each other at the table.

Fritz thought the levity made for a perfect time to talk about his move. "Well..." he said, then paused as everyone turned their attention to him. "I have some big news." Berta thought he would introduce a new word. John wondered if he would tell them about an adventure that caused his tardiness. Karl just blurted, "So what's the big news from the city?"

Fritz looked at him and stammered, "T-that's exactly what I want to tell you..." He looked at the rest of the family and said, "...you all." Silence loomed over them, but Karl made a swirling motion with his hand as if to say, Get on with it. Fritz seized the moment. "I've been offered a job in the city, and I accepted it."

Karl slammed the table, startling everyone, and exclaimed, "I knew it! You've been drooling for a city life for years."

John and Berta reacted more slowly. John finally said, "You weren't going to talk with us before you accepted?"

Fritz had his defense ready. "You know profits will be down this year. This way, you'll have one less mouth to feed." Then he hoped he was lying when he added, "It's only temporary."

Berta chimed in, "But Indy is so far away. Where will you stay?"

Karl, who knew about Fritz's romance, said, "You're not going to stay with that damned whore!" The rough language and news took Amelia by surprise. She was not offended by Karl's words; she simply didn't expect it from him.

Only Karl knew about Lizzy because he had met her once when the brothers traveled to Indianapolis on other business.

Berta looked shocked; John turned quiet as usual. With just a few words, Fritz turned their world upside down. Berta finally said, "What will you be doing, and who is this woman to you?"

Fritz folded his hands and prepared to tell them the rest of the story. He braced himself for the outcome and then said, "I'm going to be a bartender."

Karl smacked the table again. "At the Blind Pig?" he asked.

Fritz gestured with his right hand and answered, "Kind of appropriate for a farm boy, right?"

Karl retorted, "That's not the kind of pig they are referring to." The others looked puzzled, so he continued, "The blind pig refers to the police, who look the other way. Fritz will be working at an illegal speakeasy."

"Everyone looks the other way," Fritz protested. "People just want to have fun."

"Not the right crowd, Fritz," Karl responded.

Berta then asked, "Who is this woman?"

"She's the woman I want to marry."

Karl taunted, "You're not married now, but you certainly have slept with her," This news shocked Berta and Amelia. Fritz wanted to say, well, you and Dad aren't married, and you're sleeping together, but he knew better than to go there.

Bitterly, John spoke up, "Well, when the hell will you start? The sooner, the better as far as I'm concerned." No one had expected this response. He continued, "You've wanted off this farm from the beginning. But let me tell you one thing. When things get tough, and they will, you can just drink your dammed booze and starve." At that, John left the table, grabbed his coat, and walked outside. Berta thought about going after him but returned her attention to Fritz.

Karl said, "Well, Fritz, I think you just pissed off the one man who fought for everything for you."

The table went silent.

Karl drove Fritz to Indianapolis the next day, the two riding in silence. Lizzy was at work at the diner when they arrived, but Fritz knew where she kept a key hidden. Karl helped Fritz carry his bags to the apartment door, but the older brother shook his head and stared at the entry floor when Fritz invited him inside. Finally, he said, "I would be wrong if I didn't wish you good luck, but damn, you sure rocked the Hoffman boat." At that, Karl turned and walked away.

Chapter 8
June 1925

As the sun set on a warm June day in 1925, Karl, Amelia, and John sat on the back porch and enjoyed the view. It had been a long day of chores. Patrick played in a sandbox filled with all kinds of metal toys that though sturdy, often produced pinched fingers, cuts, and scratches. Patrick rarely complained.

Karl had spent the night before drinking with Moonshine Moore and Bar. They had put down several samples of a new batch of moonshine. Tonight, Karl wanted to spend time with his family, even if he quietly snuck sips of the spirit in plain metal cups. He spilled a few drops and cussed because he didn't want to waste any. He had paid Moonshine good money for this special supply that he brought home last night. John smelled the spirits and smiled. "You know you don't have to hide that and act like you are drinking water." Red-faced from both embarrassment and the sun, the man looked at Amelia. Then John laughed. "It's all right. I've had more than my share for a lifetime, and my attention is now on the Jardin."

Karl became somber and said, "I shouldn't have bought any from him, with all that he says." He frowned.

John, well aware of Moonshine's hateful words, inquired, "Well, what is he saying this time?"

Karl glanced at Amelia and then John. "You know how he is about his church and all." John nodded. "He doesn't think it's right with you and mother not remarrying."

John chuckled then said, "I'm supposed to be concerned about what Moonshine thinks?"

Karl added, "Not what he thinks, but what he tells others. He...." Karl hesitated and rubbed his hands on his coveralls. "Well, he tells others that Samuel is a git."

John queried, "A git, huh?"

"That's an Irish word for a bastard, Dad," Karl explained. "He thinks that nothing good will come of Samuel."

John quickly snapped, "I know what he says." He thought for a moment. "Let me tell you this. What Moonshine or other zealots think about my family and me is of no concern. We are proud Hoffmans. We are proud cattlemen. I love your mother, and she loves me. That's all we need." John sat silently and watched the sun set.

At that moment, Berta walked around the edge of the porch with wide eyes and said, "I will never tire of this gorgeous view." Red and yellow beams shot straight up from the horizon into the sky.

"And me neither," John answered. Berta returned John's direct gaze at her and smiled.

She joined the others on the porch and watched Patrick in the sandbox. "He's going to need a run under the well pump before he enters the house."

Karl looked at Amelia, who immediately exclaimed, "Hey, I did it the last time. It's your turn." Karl didn't argue because he enjoyed dousing his younger brother with water from the well, and Patrick liked splashing around on a hot day.

He told Berta, "I'll rinse him off before we go in."

John stared at the ground and said, "I think I've said all I need to say about Moonshine Moore." Berta looked puzzled, but she didn't push for more information, and the men didn't offer it. They knew her distaste for the man. Instead, they watched a few clouds form on the horizon as the day began to cool. The speckled clouds and brilliant sun continued to spew yellow and red rays across the prairie sky. It was a peaceful end to a long day.

A few weeks later, Berta hung laundry to dry on the clothesline as Samuel, just over a year old, slept in a stroller. She smiled and waved at Karl, who drove the Model T pickup down the lane to the farmhouse. Jackie, a neighbor's son who John had hired to help out from time to time after Fritz left, walked the big Percheron back to the barn. For some reason, the Model T backfired as it passed Jackie and the horse. In a flash, the horse jumped and pulled the guide rope through Jackie's grasping hands. Usually, he would have worn heavy leather gloves, but he had left them in the Model T that Karl drove home that day. The lead rope ripped through Jackie's grasp, burning the palm on his right hand. The Model T backfired again. The sound echoed between the house and the barn. This time, the heavy workhorse jumped a small fence and ran into the yard, kicking and neighing wildly.

Berta turned in surprise as the horse intruded into the yard. With one last loud sputter from the Model T, the horse kicked up its back legs and snagged the stroller. Immediately, Samuel burst out of the carriage and sailed fifteen feet in a high arc. Berta screamed as she helplessly watched Samuel fall and strike the first step of the wooden porch. The edge struck the baby's neck, and then he bounced off the step and twisted and bashed the back of his head into the edge of the second step. It looked as if the fall broke his jaw and crushed his throat. Berta ignored the horse and ran to Samuel. Jackie held his rope-burned hand against his body, then quickly skipped the fence to help.

Berta, falling at Samuel's feet, saw no movement nor heard a peep from the baby. She scooped him into her arms and ran inside. She turned and ordered Karl, "Go get Doc Bane now!"

She entered the house and ran into the master bedroom. Slowly sitting down, she cradled Samuel and stroked his lips to elicit a response; shock gripped her mind and body. John, who had been in the hay mound, had looked out the barn door with the first backfire of the Model T and saw the horse jump. Clumsily, he sped down the wall ladder, missed the last two rungs, and fell on his backside onto the barn's dirt floor. Jumping up, he shook it off and ran for the house. He found Berta crying hysterically in the bedroom. John never dreamed he'd use his battlefield skills again, but he found that they kicked in immediately.

Placing one hand on Berta's shoulder, he used the other to take Samuel from her. He pursed his lips, made a quiet, soothing sound, and inspected the child. Already Samuel's neck had begun to swell, and blood oozed from the back of his head. Berta began to calm herself and ran to collect bandages and water. Samuel wasn't breathing, and John assumed the swelling in his neck was choking him. He pressed his lips to Samuel's mouth and forced air into his lungs before pressing on his chest and forcing air back out. He could see Samuel's tiny chest expand and contract as he repeated the actions. After a few more attempts, Samuel spasmed with an arch in his back, sputtered, whimpered, and cried between gulping breaths of air.

Berta gasped when she returned and then laughed hysterically when she saw her son breathing again. She calmed herself immediately and attended to the wound on the back of Samuel's head, cleaning it and applying a bandage to control the bleeding. It seemed like hours before Samuel quit crying. They both tried to imagine how the infant's mind processed what happened to him. In a little under an hour, Karl brought Doc Bane into the bedroom.

At the doctor's request, Berta described the accident through broken sobs. Meanwhile, Doc pulled out his stethoscope and examined the child. He told John to grab a towel and some ice, then applied it to Samuel's swollen throat. Doc looked up at the couple and said, "If it hadn't been for your fast actions and the boy's big, healthy size, I fear he would have suffocated from the swelling of the larynx." He returned his attention to the baby and said, "We aren't out of the woods yet. He will need a twenty-four-hour observation. You need to keep the ice on his neck, even if he fusses. The next twenty-four hours may tell us what will happen."

Five-year-old Patrick had followed Berta into the bedroom. When he saw his baby brother's limp body, white as the sheets on the clothesline, he ran from the room and hid under the kitchen table. He pulled his knees up to his chest, grasped them, and rocked from side to side.

After a time, Amelia noticed Patrick under the table. She smacked Karl on the shoulder, gestured toward the young boy, and whispered, "Well, do something!"

Slowly, Karl approached his brother and sat on the floor next to the table. He reached out to touch Patrick, but the boy leaned away and gave him a frozen glare.

"Patrick, he'll be alright," Karl muttered, unsure of himself. He glanced toward the bedroom and then back to Patrick. "You will be all right," he said. "We are all going to be all right."

For the rest of the day, Patrick avoided everyone and remained under the table. Nearing nighttime, he crawled out and put himself to bed. Karl and John closed down the farm for the night, did the nightly animal chores, and parked the Model T in the barn. As Karl took Jackie home, the young man nursed his rope burn, which Amelia had treated. "That's going to burn for a while," Karl said.

"I don't care. I deserve it," Jackie replied mournfully.

Karl caught on quickly. "There's no way you could have held that huge animal down by yourself. It's a fluke," he reassured Jackie.

The young man sat quietly and tried to think of what he could have done differently.

John and Berta took turns cradling Samuel and lightly holding an ice pack to his throat. Samuel was expressionless but breathed regularly. Within hours the swelling began to recede in his throat. Between prayers, Berta continued to check his bandaged head. When the cut was bloody, she cleaned it, thankful that the back of the head didn't swell.

The next afternoon, Doc rolled down the driveway with his horse and buggy. Karl kept up with the chores with Amelia's help, occasionally pausing to check on Samuel. They tended to the cattle and retreated to the fields. As much as they wanted to be inside with Samuel, they knew the alfalfa was ready to be baled, and a hay failure could translate into a cattle farm failure over the harsh winters. Haymaking demanded precise timing because too much sun, rain, or dew could damage the crop once it was cut. Karl knew they must gather the alfalfa and stack the bales in the barn. So, as they worked, the two took comfort in the knowledge that Doc was very, very good with babies. Karl recalled the time he saved the Danbury's baby when he was run over by a horse. Farms were dangerous places for young and old alike.

After Doc attended to Samuel, he placed him in the center of the master bed. John and Berta had sat on the bed all night, not even turning back the bedsheets. Slowly, Doc turned from the child and wiped his brow with the back of his shirt sleeve. "This child is the strongest I've ever seen," he declared. "The fact that he survived the impact is testimony to his strength." He hesitated, then continued, "On the good news side, his skull seems to be okay. He should recover from that bump just fine."

John and Berta studied Doc's every word, like hearing a judge's sentencing of a criminal. Berta reached out and touched Doc on the forearm. "But?" she asked with a grimace on her face.

Doc continued, "I'm afraid his larynx is crushed."

Berta let out a crying gasp. "Meaning..."

Doc paused, stared at the floor for a moment, and shook his head. When he returned his look to the couple, he said, "Samuel may never be able to talk."

John, with his arm around Berta, tightened his embrace. Berta put a hand to her face, then turned and saw Patrick standing at the door. He still wore his pajamas. A little bit of cereal clung to the front of his pajama shirt.

Patrick slowly walked past John, Berta, and Doc and climbed onto the bed. He crawled over to his baby brother and stared at him for a moment, then reached out and gently took Samuel's hand. He sat beside the child, who had been so quiet since he quit crying yesterday. Samuel slowly turned his head and looked at Patrick. A smile widened on his lips. Patrick patted Samuel's tiny hand and said, "You are going to be all right." He kept patting his hand. "We're all going to be all right."

Doc smiled, too, and mused, "Well, I've never seen that before. I think Samuel's going to have some help along the way."

As Patrick began to talk with Samuel, asking if he wanted a drink, Doc continued, "There's a new hospital that just opened in Indianapolis called Riley Hospital. It specializes in care for children. I will come back tomorrow and check on his progress, but I suggest you prepare to take Samuel there for some expert opinions. It's an option. Think about it. And, in the meantime, you two take shifts and get some rest. I don't need to attend to you for symptoms of exhaustion." He patted John on the shoulder, left the house, and strolled back to his horse and buggy.

John and Berta didn't need to think twice about whether to take Samuel to Indianapolis. Karl had proved, summer after summer, that he could attend to the hay harvest. "Hire some extra help if you need," John told his son. "Don't worry about the cost."

And with that, John and Berta began to prepare for the trip to Indianapolis. Doc did as he promised. The next morning, he stopped by to look upon Samuel. By now, Patrick had become inseparable from his baby brother. Doc wrote a referral to admit Samuel to Riley. As he departed, he said, "Be sure to take Patrick with you. Something is going on there that's stronger than medicine."

The ride in the Model T took two hours, with Berta holding Samuel the entire way. Patrick sat between John and Berta and frequently held onto Samuel's foot as the pickup bounced along the roads. John and Berta were awestruck as they entered the hospital. Everything from the engineering in this new building to the professionalism of the entire staff impressed them. Patrick was allowed to remain with his brother except when they performed an X-ray on his throat. After about two hours, the physician joined them in a consulting room. He apologized for the delay and explained, "It took ninety minutes for the X-ray to develop."

John piped up, "Is there anything you can do?"

The physician nodded in acknowledgment, then replied, "I'm afraid the question is, what can you do? Your local doctor's referral is accurate, but the damage is permanent. It's what you do from here that will make a difference for Samuel." The doctor paused and smiled. He brushed his hand through Patrick's hair. "This one already knows what to do. Samuel will need your care. His struggle to communicate his needs will be with him and you forever."

The doctor explained that speech therapy would be of no help. "Sign language will be a key to communication," he explained. "We can give you some medicines to help the skull abrasion and the throat to minimize further issues, but again, this young man will never be able to voice his concerns. I hear Lafayette has some experts who can help with the sign language aspect. I can make some recommendations. In the meantime, we'd like to keep him overnight and make further observations." He looked at John and Berta, "Will that be a problem for you?"

John looked at Berta and then nodded to indicate that would be fine.

 "You can stay as late as you want," the doctor said, "but I recommend you arrange for overnight accommodations. The front desk staff developed a list of places to stay nearby. They can share it with you." He noticed the exhaustion on their faces. "Our observations are to see if we missed anything. I feel certain that young Samuel's skull abrasion will heal nicely. His strength obviously helps him in this battle. I'm sure he will be a great help someday on your cattle farm. In the meantime, my advice is to get some good rest tonight."

The Hoffmans remained by Samuel's side until around 11 p.m. The chairs in the room were uncomfortable, to say the least, and John rubbed his aching back. Finally, with Patrick asleep in John's arms, they walked to the hotel.

The Hoffmans returned to the hospital early the next morning after grabbing one of their first full meals since the accident. Riley Hospital released Samuel and gave John and Berta a list of referrals in Lafayette.

Driving down the long lane of the farm, the couple saw Karl and Amelia maneuvering through the pasture while cutting the alfalfa. Berta quickly made lemonade and sandwiches, and John took them out to the pasture. He apprised them of Samuel's situation, and they, in turn, informed him that the cutting was going great. John said, "Well, I'm back and here to help. Amelia, get some rest and help Berta as much as you can."

Slowly, the everyday farm life returned. In the meantime, Patrick doted on Samuel and helped him better than any five-year-old could.

Chapter 9

June 1925

All John remembered from the days after their return from the hospital was that he and Karl took wrecking bars to the Model T that backfired and ultimately resulted in Samuel being thrown against the steps. They smashed the Model T, striking it again and again, breaking windows, crushing doors, and cutting tires with knives. Eventually, the Model T crumbled into a heap of metal, rubber, and glass. Then, with a jolt, John woke from his dream in a sweat. It had been a week since the accident. His heart raced, he breathed heavily. Berta woke and placed a soothing hand on his arm. He hated the car for what it had caused Samuel. In the middle of the night, with his heart racing, he decided he would get rid of both the car and the horse.

The next day John drove into Lafayette and purchased a Dodge Brothers one-and-a-half-ton black pickup. A four-cylinder engine powered the new truck, which outclassed the cobbled together Model T. One of the newest factory-built pickups, it sported an electric starter and a built-in heater. John paid $490 for it and told the dealer to deliver the vehicle to his farm after they prepared it. He drove the Model T to Doc's place and donated it to him. Karl had brought the Hoffman's horse and buggy into Boulton and waited at the soda shop, devilishly enjoying the best chocolate soda in the county. John smiled when he walked in, looked at the shop owner, and said, "I'll have what he's having."

Karl watched his dad pull out a chair, the metal feet scraping loudly against the floor. He asked, "Have you been to the butcher?"

"Yep."

"When are they going to pick up the horse?"

"Tomorrow," John said quietly. "We will be down to two horses after that." It bothered John to sell the horse to the butcher, but he couldn't deal with its nightmares. He asked him to donate the meat to needy people.

The men rode home in the buggy, alone with their thoughts. Karl broke the silence. "How do you think Fritz is doing?"

John glanced at Karl and then gave the reigns a pull and stopped the horse and buggy on the road. "I was just thinking about him. Why don't you take the new Dodge down to Indianapolis and check on him next week?"

Karl nodded. "I'd like to do that." Fritz was on the minds of both men after Samuel's accident.

The new Dodge was delivered the next day, and Karl ran his hands over the pickup hood. "This is one big and powerful machine!" he exclaimed. Berta and Amelia both stood looking at the new vehicle. They didn't realize it at the time, but this would be the first in a long tradition of Hoffman family inspections of new vehicles.

Three days later, Karl went to Indianapolis to check on Fritz. At Lizzy's apartment, the three of them chatted for more than an hour. Finally, Karl related the story of Samuel's accident. "It happened so quickly," he said. "I drove into the farmstead and waved at Mother hanging laundry while Jackie led the horse to the barn. That's when the Model T backfired. I have no idea why. It was some sort of weird fate."

Fritz was stunned by the news of what had happened to Samuel. He sat shocked not only because of the accident, but he thought about the fact that he was no longer a main part of the Hoffman family. Finally, Lizzy broke the silence and asked, "How is Samuel now?"

"The doctors say he'll grow up fine, but he will never speak."

Fritz gulped and whispered, "That's a shame."

Karl changed the subject. "How are you two doing?"

Fritz reached out and hugged Lizzy and said, "We're doing great, right Lizzy?"

Lizzy half smiled and answered, "Just peachy. Which reminds me, do either of you want a sip of peach Schnapps?"

Karl observed Lizzy. She was stunning, but he also noticed that her answer was not sincere. He dismissed that thought and looked at the clock. "It's two o'clock in the afternoon. A little early for that, isn't it?"

Fritz jumped up anyway and went to the shelf in the apartment's kitchen and then took down a bottle of Schnapps. "It isn't as good as Bar Moonshine, but I've sold all of ours to the Blind Pig. If I knew you were coming, I'd have asked you to bring a few cases," he told his brother.

Karl replied, "I'll let Bar know you are out. I'm sure he will take care of you."

Fritz put three glasses on the table, poured them full, and slid one each to Lizzy and Karl. Lifting his glass into the air and nodding to the others, Fritz said, "Here's to good times."

Karl left the apartment an hour later, departed Indianapolis, and headed home. He spent the entire drive home trying to figure out what was wrong with Fritz's relationship with Lizzy. Something was off, but he couldn't put his finger on it.

Meanwhile, Samuel continued to recover, with Patrick hovering over him. That morning, a milk bottle and a bowl of oatmeal stood on a bedside table. Instinctively, Patrick looked at Samuel, then held up the milk bottle and pointed at it with a single finger. Samuel shook his head to indicate no. Patrick set the bottle down and picked up the oatmeal, and pointed to it with two fingers. Samuel nodded yes. Patrick sat Samuel up against a pillow and began to spoon-feed him. Just then, Berta entered the room and stopped in her tracks. She turned toward the door, put a finger to her lips, and motioned for Amelia to come and see. The two women smiled as they watched Patrick and Samuel interact. After another spoonful of oatmeal, Patrick turned and saw his mom and Amelia watching him. He tilted his head matter of factly and then returned his attention to his brother, who now held up one hand with his index finger extended. Without hesitation, Patrick set the oatmeal down and picked up the milk bottle. For years this interaction would be repeated over and over, growing more and more complex.

For more than a year, Fritz thrived in Indianapolis. His bartending and storytelling skills made him very successful. The moonshine that he brought from Moonshine Moore's farm drew customers to the Blind Pig in packs to taste the spirit everyone said was great.

Lizzy also benefited from the increased business; she received generous tips like Fritz. She bought a short outfit that looked like it belonged in a burlesque show. The black dress hung on her narrow shoulders by two straps, and the hem fell just below her hips with a very light fringe that hung to her knees and bounced back and forth as she walked from table to table. She became the talk of the town. Not only did Lizzy seem not to care that some men placed their hands on her backside and ran their fingers down her lean legs, she seemed to encourage the behavior from some. Fritz noticed, but his enthusiasm never dimmed. Waking every day around noon, he rushed out to buy copies of the Chicago City Times and the Indianapolis

Independent. The knowledge of the two cities fueled his banter in the evenings.

Brad, the speakeasy owner, couldn't have been happier that business was roaring. What he failed to realize was that the business drew unwanted attention.

Months after Fritz and Lizzy brought their talents to the Blind Pig, five well-dressed strangers entered the speakeasy. Seated at their own table, one man paid close attention to Lizzy. When the place was jumping, he discreetly slipped a note into Lizzy's hand with a twenty-dollar bill attached. After two drinks, the men unceremoniously left the Blind Pig. That night Lizzy left early, telling Fritz that she wasn't feeling well. Fritz patted her on the head, kissed her good night, and continued talking to others at the bar.

Once outside, Lizzy approached a brand-new Cadillac Town Sedan. The driver hopped out and opened the back door for her. She climbed in and asked, "Where to now?"

"To a place out in the country," said the stranger who had given her the note and the $20.

Fritz later closed the bar and was shocked when he returned home to find the apartment empty. Frantically, he searched fruitlessly for a note. Finding nothing, he sat on the bed with his back against the headboard, knees tucked under his chin, and his arms around his legs. He remained that way for hours, sometimes rocking back and forth. The sun rose before Lizzy opened the apartment door. Fritz jumped up and nearly fell because his legs had grown numb. Lizzy smiled then said, "I want you out of my apartment."

Shocked, Fritz gasped. "Wh-…" He couldn't even frame a question.

Lizzy moved about the apartment and gathered a few things.

Finally, Fritz sputtered, "W-w-where have you been?"

"None of your business!" Then she continued, "I've given my key to the landlord." She walked to the door with Fritz following. She turned and said. "I'm leaving this morning for Chicago. I want you out in a week. Turn your key into the landlord. I've collected what will be the unused portion of the month's rent. You can keep the deposit."

"B-but, why?" Fritz asked, wincing.

"I got a better offer from someone other than a mere farm boy turned bartender," she answered, her brutally honest words stinging Fritz. With that, she turned and walked out the door. Fritz stood motionless for a few moments.

That evening he trod slowly back to the Blind Pig. Everyone asked about Lizzy, with some men more curious than others. Fritz just shrugged and said, "She ran off to Chicago."

Losing Lizzy was just the start of a bad day for Fritz.

Well past midnight, near closing time, four of the well-dressed strangers returned to the Blind Pig. Only Fritz, Brad, and one customer remained. This time, when the strangers walked in, two of them carried Tommy guns, and two carried sticks. One man fired a short round into the back of the room, and the two with sticks began smashing bottles of moonshine. When Brad resisted, they bashed him in the face and dropped him to his knees.

Bloodied, the veins in his neck bulging, he cried out, "What's going on?" The four men stood over Brad as the last customer ran for the door. Once he exited, they shot Brad in the head. Then they turned their attention to Fritz.

One of the men spoke: "I'm in charge now." He looked Fritz over, then added calmly, "I heard you lost your lady. Chicks like that come and go." Fritz studied the stranger and tried to understand what had just happened. The stranger continued, "Relax, kid. We're not here for you. In fact, we like the way you work. You can stay on under the new management." With that, the two men with the clubs picked up Brad's body and carried it out the back door. The lead man said, "We'll take care of him. You stay and clean this mess up. We'll be back tomorrow with Chicago-style hooch."

Fritz didn't know what to think, but dutifully, he began to clean up the mess. About 4 a.m., he closed the outer door to the bar and walked back to the apartment to another sleepless night. At 10 a.m., he grabbed his key to the apartment and stuck it in an envelope with a note to the landlord that said: "Lizzy ran off, and I'm leaving too. Keep the deposit. We are both gone." With that, he grabbed his heavy gear bag and left.

He walked to the train station and purchased a one-way ticket to Boulton. When the train arrived there two hours later, he started the walk home. It took Fritz almost five hours to cover the nearly fifteen miles. A couple driving down the road offered him a ride, but he courteously declined. Fritz wanted the time to think, to clear his head, and prepare for this return home with his tail tucked between his legs. He thought that John might reject him. What would he do then? He thought about Karl mocking him and saying, "I told you so."

The walk began to take its toll on Fritz; he carried his heavy luggage over his shoulder. He was out of shape and had underestimated the distance. At one point, he stepped on the edge of the road and lightly sprained an ankle. He considered it a punishment that he deserved.

Later that afternoon, John sat on the porch and watched Fritz walk up the driveway. He called for Berta. When Fritz approached the house, his parents could see that he was in shock. Instinctively, John recognized the look of a man who he had witnessed a murder. Without hesitation, he approached his son and hugged him. Then he surprised Fritz by saying, "I'm sorry for the way I spoke to you the night before you left." He hugged Fritz tighter, then backed away slightly, keeping his hands on the young man's shoulders. "Come inside. You look as if you could use some coffee." John never asked Fritz what he had witnessed. He would let Fritz decide when to open up about it.

With Fritz home, John and Berta decided to remarry that summer. They kept the ceremony simple and held it on the porch of their house. John, Berta, Amelia, Karl, Fritz, Patrick, and Samuel made up most of the wedding party. In addition to the preacher, Berta invited a woman from Boulton to be the witness. John smiled when she told him who she wanted to serve as the witness. "Woman, you never cease to surprise me. You invite the biggest gossip in Boulton. You want everyone in town to know we remarried, don't you?"

Berta said nothing but returned John's smile.

Chapter 10

Christmas Eve 1928

Two and a half years had passed since Samuel's accident. 1928 was winding down, and the Hoffmans faced a blizzard on Christmas Eve. Living on a cattle farm during a blizzard tested every farmer's will. Just because it was cold outside didn't mean they could stay inside by the fire. Much needed to be done this Christmas Eve to care for the cattle. John, Karl, and Fritz struggled against the blowing, stinging snow. It struck their faces like cutting thorns and turned the cattle white with snow and ice. Karl's mustache, which he grew for John and Berta's second wedding, turned from light brown when he woke that morning to white and sticky with snow and ice. November and most of December provided colder than normal temperatures but not as cold as this last week before Christmas. In one way, that had been a good thing because the cattle had developed heavy winter coats that helped them fight off the cold when temperatures plummeted even further during the past week. Even the two Percheron, enormous workhorses who typically preferred not to be cooped up in the barn, gladly came in from the elements and entered their stalls.

The biting cold and the wind meant every chore took twice the effort, twice the energy, and twice the time. During the winter, the Hoffmans fed their cattle late in the day, so the high point of digestion would occur overnight when temperatures were at their lowest. That advantage, along with the Gelbvieh breed's quick maturation, would help them survive the blizzard. That knowledge helped calm John, who was gravely concerned that they might lose a few cows.

During the summer, the Hoffmans had separated the high-quality hay from the average hay, and John ordered the boys to fill the outdoor racks with the best forage. They finished by dumping feed supplements into the cattle troughs. With the right combination of care, the cattle would fight off the cold and maintain or possibly improve body condition. The care provided by the men meant contented cattle and higher profits at the market. The chores ate up four hours that afternoon, taking them into a darkening December sky.

The house felt like a blast furnace when the men returned. Even so, it took a half-hour for the ice to completely clear from Karl's mustache. The three men stopped in their tracks when they entered the kitchen, taking in the wonderful smells. The aromas of hot breads, freshly baked pies, and a roasted turkey wafted toward them. Berta and Amelia had prepared considerably more food than they needed so they wouldn't have to cook a big meal on Christmas day. Taking in the scents, the men stomped the snow and ice off their boots and hung up their winter gear on the kitchen hallway rack. As soon as Karl hung up his things, he quickly stepped to a kitchen counter, grabbed a roll, and stuffed it into his mouth.

"That's for later!" Amelia warned him. Although she admonished him, Amelia smiled with pride that something so simple could soothe a beast of a man like Karl, who had just returned from the frozen country air.

John grabbed Berta's backside, eliciting a playful smack to his cheek. With his face still frozen, the slap stung slightly. Berta noticed his painful grimace, then shook a finger at him and said, "That will teach you!"

The family had purchased a radio in April, and this Christmas, they listened to holiday music playing on Chicago's first 50,000-watt radio station. The radio had miraculously transformed the Hoffman home, bringing them the Chicago Cubs in the summer and the Bears in the fall. But Christmas brought music over the airwaves. Both fair of voice, Berta and Amelia sometimes sang along with the hymns, belting out soaring melodies and finding soft tones for the mellow notes. The men enjoyed and listened to them add to the songs. The house was suddenly filled with singing, and Berta and Amelia outdid themselves. When Coventry Carol played on the radio, John retreated to his comfortable chaise lounge in the back of the family room. He listened closely, especially when the women softly sang the last words of the song:

That woe is me, poor child, for thee

And ever mourn and may

For thy parting neither say nor sing,

"Bye bye, lully, lullay."

John thought about the story of King Herod ordering the death of newborn male babies and the Massacre of the Innocents. Berta and Amelia's voices brought the story to life. John rested in his chair, his eyes welling with tears. He discreetly blotted them away, unnoticed by anyone else. When the song was over, the women were silent in the kitchen and reflected on the horrible actions of such an ancient time.

The announcer returned to the air and said, "Sorry, I had to play this centuries-old song, but let's move on to something happier. Here's the Chicago symphony playing Joy to the World.

At that moment, Amelia announced, "It's time to eat."

Karl and Fritz rushed to the table, their stomachs growling from the long day of work and the enticing aromas spilling from the kitchen. John trailed them, pausing to turn off the radio. He entered the kitchen and said, "Let's take time to say a prayer and be thankful for all that we have." Afterward, standing by their chairs, each professed their gratitude. John started: "I am thankful for the talented women who entertained us after we worked hard all day."

Berta smiled and said, "I'm thankful I didn't have to work out in the cold today."

Karl chimed in, "Why don't you go next, Amelia?"

She didn't hesitate and said, "I'm thankful for this family, taking me in and making me feel so much a part of it."

"I'm thankful for the new radio and the world that it brings to our house," Karl stated.

Fritz stomped his foot and exclaimed, "That's what I wanted to say!" He looked around and finally said, "I'm thankful for this farm that provides food, shelter, and a wonderful life." John raised an eyebrow but didn't say a word. In that brief moment, father and son communicated more than they had at any other time.

They all turned to Patrick, who said, "I am thankful for Samuel because he means I'm not the youngest." Patrick used his hands to signal to Samuel, who was now five years old. Samuel signed back to Patrick, who interpreted for the others. "Well, I guess Samuel is thankful for water!" The family laughed, taking Patrick aback, but Samuel just smiled.

The young brothers had developed a complicated method of communicating. Patrick read books and stories to Samuel when he was younger, but Samuel began to read simple books on his own in the last month. Having learned the alphabet, Samuel communicated with Patrick with amazing speed, especially given the complexity of their process. Patrick and Samuel had assigned numbers to every letter in the alphabet. At some point, Patrick drew a chart matching the letters and numbers.

For A, Samuel held up one finger. Ten fingers represented J. The rest of the alphabet became more difficult, but the two youngest boys worked out an efficient system once memorized. It involved the left hand pointing at the right hand for 11, 12, 13, 14, and 15, which was k through o. Then they rotated, pointing at the other hand with one or two fingers until they reached the 26th letter.

Berta watched Patrick and Samuel and marveled at the system they had developed. "I still can't read it, but sometimes, it appears that Samuel doesn't spell out certain words," she observed.

As they sat down at the table and began to pass the food, Patrick said, "That's right, and I can explain. It's all about context." Berta smiled because that seemed like a big word for a nine-year-old. "I usually know what Samuel will say," Patrick explained. Samuel smiled and nodded. "So, once I get a few letters of a word, I nod that I know the word, and he moves on to the next word. And, being five years old, sometimes he misspells a word. I usually correct that in my head." Samuel shrugged. "With common phrases, he sometimes only gives me the first letter of the words," Patrick said. Then he demonstrated: "For 'good night,' he just indicates G, the seventh letter in the alphabet, and N, the fourteenth letter."

At that, Samuel held up seven fingers.

"G," Patrick said.

Then Samuel pointed one finger on his left at his right hand on which he extended four fingers and made the sign for N.

"N," Patrick translated. "That G and N, at the appropriate time of day, stand for Good night." Samuel smiled again and nodded.

Karl watched them, his chin resting in his hand, and considered the two boys' determination to develop the process. "It's a good thing the Hoffman mind knows math so well," he said. "Now, let's eat." Then he smiled and winked at Amelia.

Berta caught the wink and said, "All right, you two. Do you have something you want to tell us?"

They both raised their eyebrows at Berta's comment. Then they looked at each other. Amelia nodded as if to say, Go ahead.

Karl scooted his chair back, stood, reached out to grab Amelia's extended hand, and guided her to stand with him. She nervously brushed her dress with the other hand as if to knock off any breadcrumbs. Karl started, "Christmas Eve is as good as any time to tell you. Amelia and I are engaged, and we plan to marry in early April. With this weather, we haven't gone into Lafayette to get an engagement ring."

"That's wonderful!" Berta exclaimed with her hands to her face. The rest of the family applauded the news.

"Hear, hear!" John chimed in enthusiastically.

Fritz half laughed and blurted out, "It's about time." Karl reached around Amelia and popped Fritz in the back of the head. Karl and Amelia bowed and then sat back down and resumed their Christmas Eve meal. For the rest of the dinner, the conversation centered on the wedding and impending changes.

"Fritz, I hope you don't mind giving up the rental house for the two of us," Karl said.

Fritz smiled and then replied, "Gee, you don't want a third wheel around? Of course, I don't mind moving back into this house."

Then Karl directed his attention to John and Berta and almost pleaded, "I hope that's okay with you two."

Berta placed a hand on John's arm and answered, "That will be a good arrangement. Don't you think, John?" He nodded in agreement. She turned back to Karl and Amelia. "Have you picked a date?" she asked.

Amelia piped in, "We think, Sunday, April 7. We don't want a big wedding, just the local pastor at the church and a few others." She hesitated as she realized this family was the only family she had. The thought that she would be the only member of her own family at the wedding caused her to feel alone for a moment. Karl sensed something was wrong. He put his hand on her shoulder and asked for her thoughts. She answered, "I don't know what overcame me, but suddenly, I miss my family back home in Bavaria. I won't have any of my family at my wedding."

"You'll have us," Berta said, "We are your new family, and we will be even more so on April 7th."

Amelia smiled and considered this thought, which had only now struck her. Karl reached an arm around her and hugged her from the side.

"Thank you all," she said. "It will be all right. We will have a wonderful wedding."

John had been thinking while the others bantered. "Have you thought about what other changes will happen?" he asked.

"I'll stay on with the farm," Karl said.

John looked at Karl as if to say, 'Of course.'

Amelia piped in, "I hope that I can stay on as the housemaid." As she looked at Berta, Patrick, and Samuel, she realized she would not be sleeping in this house with them. "I guess I'll have a two-mile commute from the rental house." She looked back at Karl and said, "We'll have to think about transportation."

"That won't be a problem," he replied. "I've saved some money up. We can buy a Woody depot hack." The depot hack could seat several people and literally had wooden sides. Since 1910, they were mainly used around train depots to pick up passengers and their luggage. Before automobiles, hackney carriages had provided most transportation from the rail station. Karl liked the idea of owning a Woody, even though it was unconventional for a farm.

Fritz chimed in, "Going to start a family so soon?"

"None of your business, brother. I just thought a wagon would be more useful for all of us," Karl retorted.

Berta remarked, "I haven't worn it since John, and I remarried, so Amelia, you can use my wedding dress. It'll need a few alternations." Berta looked at Amelia's bustier bosom. Amelia blushed again but nodded as if to say, Thank you.

John sized up Karl, who stood three inches taller and certainly had a bigger build. He said to his son, "You won't be able to use my suit. We'll head into Lafayette when the weather turns, and I'll help you pick out a new suit."

Berta remembered that it was ninety-four degrees on the day they remarried. She poked at John and asked, "Do you recall how much you sweat that day?"

Suddenly, Patrick spoke up. "Samuel has something to say," he said, then looked at Samuel, who began to signal. Patrick quickly understood. "He says he will miss you, Amelia." They were silent as Samuel continued to signal. "He won't miss you as much, Karl," he said to laughter.

Berta reassured Samuel, "They will both be around here almost daily."

Fritz had snuck in to switch the radio back on, and now Jingle Bells played in the background. Immediately, Berta and Amelia began to sing along, and they were soon followed by Karl, Fritz, and Patrick, while Samuel lightly tapped percussions on the kitchen table.

The marriage would be a significant change for the family. By the next morning, the storm had subsided. Waking before sunrise, John and Berta wished each other a merry Christmas and then had a cup of coffee at the kitchen table. As the sun burst over the horizon, John donned warm clothes and walked outside to check on the cattle. The morning sky burned a brilliant red with patches of fluffy white clouds. The snow that crunched under John's steps was pink with the reflection of the morning sky. John looked over the cowherd, which had gathered together on the east side of the barn, out of the wind. Miraculously, they appeared to be in fine shape. Later, the boys would gather and go through the routine of feeding cattle. On this Christmas day, their efforts would be considerably easier than they had been the day before.

John turned and walked to the Jardin. The limbs of the cherry trees stabbed the red sky like crooked, black fingers. The evergreens crisscrossed in the same pattern as the original French Jardin Botanique. They smelled of Christmas trees in the calm morning as John walked down a path in the Jardin. The leafless rose bushes stood a few feet above the snow. Along with the cherry trees, they reminded John that life moved on in cycles. He thought, 'I'll be forty-four years old when my oldest son will marry.' He wondered how long they would take to make him and Berta grandparents. He took in a breath of fresh, cold morning air and continued to walk through the Jardin, happy for his son and Amelia. Inside the warm house, Berta watched John walk toward the sunrise. She could read his thoughts. She knew how sentimental he could be. Berta sipped her coffee and smiled.

Chapter 11

April 7, 1929

A storm loomed over northwest Indiana on April 7, 1929. Karl and Amelia had stuck to their plans to have a small wedding. Karl woke early that morning and helped Fritz and John with the chores, even though both protested that he should take the day off. Nevertheless, Karl refused to bail on his duties. The cattle had been fed early, and Berta and Amelia, with the help of a neighbor, had prepared a feast for the wedding.

The storm let loose and darkened the skies at 2 p.m. As the three men ran into the house soaking wet, Karl asked, "Why not have a gully washer on our wedding day, hon?"

"Don't think that will allow you to postpone this wedding!" Amelia retorted. Then she whispered into his ear, "I've been waiting too long for you know what." Karl turned red, and Berta, who had overheard, smiled slightly. The men left for the rental house to get dressed in their finery, and Berta helped Amelia into her beautiful dress. Leaving behind a neighbor to finish the wedding dinner preparations, Amelia, Berta, and the younger boys departed for the church.

Karl, Fritz, and John awaited them at the church doorstep, out of the rain. Patrick, now nine, jumped out of the station wagon to hold the door open for Amelia. Karl froze at the first glimpse of his bride. It wasn't just the dress; she wore her long hair in a stunning French braid. Starting at the crown of her head and weaving its way down to her long, slender neck and shoulders, this braid that Berta had fashioned sharply accentuated Amelia's beautiful features. Karl quickly moved the expression of shock from his face to a big smile. He hoped that Amelia had not noticed the shocked look. The rain poured down as Karl skipped his way toward Amelia with an umbrella. Just before he reached her, Karl slipped in the soggy grass and fell on his backside. Recovering quickly, he jumped up and held the umbrella over Amelia's head. Giggling, she started to push it away, looked at his soaked pants, and said, "You might need this more." Patrick, Samuel, and Fritz all laughed.

Samuel signaled to Patrick, who interpreted, "He said, 'Soakie britches.'" They all chuckled, and the family moved into the chapel for the wedding.

The storm let up by 6 p.m., but not before it made a muddy mess of the farm. The road approaching the farm had been made slippery by the rain, but that didn't stop neighbors from taking their Sunday drives past the Jardin. This evening, most of the Hoffmans ignored the cars passing by as they feasted on the wedding dinner.

Only Fritz's attention occasionally diverted from the celebration. Since abandoning the Blind Pig and leaving Indianapolis altogether, he often wondered if he would ever see a black Cadillac approach the farm. He hoped he was so insignificant the gangsters would not track him down.

 "Hey, Fritz, pass me the peas," Karl called out, noticing his brother staring out the window. Berta and Amelia had fixed sugar-ham, peas, mashed sweet potatoes, cornbread, and, of course, a wedding cake. Karl still had a bit of icing on his cheek where Amelia had mushed some on his face. They finished off the dinner with homemade ice cream and peaches they had canned the year before.

John stood and clinked his fork against a glass. He garnered everyone's attention and said, "I'd like to make a toast. We have enjoyed these two before, but now!" He repeated himself for emphasis. "Now that they are a couple, we will enjoy them more, I'm sure. May you be even more blessed than Berta and I have been. May you live long and happily together." Then he said quietly, "And bring us many grandchildren."

Berta smiled, raised her lemonade glass, and exclaimed, "Hear, hear! Cheers!" She thought her quiet-natured husband had given a lovely toast.

The brand-new couple remained at the farmhouse for an hour after the dinner but finally walked to the station wagon, followed by the rest of the family, who waved goodbye. They drove to the rental farm.

Suddenly, Samuel signaled to Patrick, who interpreted, "Now we can call it Karl and Amelia's place instead of calling it the rental house."

Fritz smiled and mumbled, "More like the love nest, I think."

The Jardin attracted more and more people since its completion, drawing passersby every weekend. Some even stopped, climbed out of their cars, and walked into the garden. Torn between the intrusion and the desire to share the beautiful space, John sometimes just sat in the house and let people visit; at other times, he stepped outside and talked with them. John was trimming the shrubs one Wednesday in July when a man drove his car right up to John. He stopped trimming and watched the man climb out of the car. Berta happened to look out the window. The man was well dressed and drove a very nice car. Her curiosity piqued, she put down her kitchen towel and walked outside to join John. When she approached, John turned to her and introduced the stranger. "Honey, this is Jason Schmidt from Indianapolis. Turns out that he's from the..."

Mr. Schmidt interrupted, "The Canton Foundation, ma'am." He tipped his hat at Berta.

John nodded and finished the introduction. "This is my wife, Berta Hoffman." Schmidt handed John and Berta each a card. John continued, "I told Mr. Schmidt that this Jardin is a replica of the Jardin Botanique in France. But he already knew that. It appears that Mr. Schmidt's family is also from that part of France. How did you learn about this garden?"

"You can call me Jason," he said politely. "You see, I have a painting of this Jardin on my office wall in Indianapolis. One day, a co-worker who had been raised in Boulton came into my office and asked if my painting was made from this Jardin." He paused and chuckled. "I informed him that it had been made before the war destroyed it in France. Then he told me about yours."

At that, John cleared his throat. Berta put her hand on his shoulder.

"Did I say something wrong?" Schmidt asked.

"It's a long story," John said. He turned to Berta. "Do we have some fresh coffee brewed?"

She answered, "Why don't you come up to the house, Mr. Sc…Jason, and we'll fill you in on the details."

The man nodded but said, "I'd like to stand here for a moment longer if you please."

John nodded, "Bring your car up to the house when you are ready."

Over coffee, John described what had taken place, down to every detail. "So, you see, I am responsible. I destroyed that beautiful Jardin."

Schmidt shook his head and answered, "You have quite a story, and yet you are responsible for rebuilding it in exquisite detail, according to the painting on my office wall. And you built it here on your farm."

Berta chimed in, "He is an amazing man. He did it all from memory."

Schmidt asked, "It must be very expensive to maintain, no?"

John nodded. "Yes, in both time and money."

"I'm not here just to admire it," Schmidt said. "I'd like to offer help."

Puzzled, John reflected on this interesting offer. He'd never considered such a thing. The Jardin was his life's dream and his life's penance. He had traveled from war to alcoholism to this dream that others admired. Finally, he said, "Help how?"

"Well, I am open, and my foundation is open."

John replied slowly, "It's obviously my dream and, maybe not so noticeable, my penance for destroying the original in the war." Schmidt began to dismiss the actions of a war hero but hesitated. John continued, "While I am flattered that you are interested, I'm not ready to share it that way." John felt greedy and flustered. Schmidt could see that.

"Well, you have my card. If there's anything I can do to help, please contact me. In the meantime, do you mind if I walk through it again?"

John motioned with one hand toward the garden and said, "If you don't mind me tending to the shrubs."

That September, the Hoffmans began to hear reports of strange activity in the economy, and the stock market in particular. The economy, up to August, had been booming. But now, the continuous broadcasts about the economy bled into October. Most had thought the boom would never end, but the news broadcasts turned darker each day. John, however, thought to himself, It was bound to happen. Though the Hoffmans were comfortable financially, they had never invested in the stock market nor dabbled in the Chicago Board of Trade. On October 24, 1929, the biggest crash in the stock market occurred. The next day, Friday, they sat around the kitchen table and listened to reports of the market failure. They heard reports that people had jumped out of high-rise buildings.

"They deserve it," John said.

Berta smacked him hard on the shoulder. He continued, "Well, they are all a bunch of greedy crooks who wanted to get rich quick. Real estate is the only place to put investment money."

Berta continued to admonish him, "Still, you shouldn't feel that way about people you don't know."

"I don't know them," he allowed. "They don't know me. But they may have ruined our agricultural markets and my real estate investments in Indianapolis."

The Hoffman family continued to follow the news; the newscasters referred to October 24 as Black Thursday. One investor, interviewed by an Associated Press reporter, compared the 1929 stock market crash to Black Friday of 1869, explaining that, in that case, a few greedy investors drove up the price of gold. U.S. President Ulysses S. Grant dumped U.S. gold on the market to offset those trying to drive up gold prices. His actions crashed the gold market and eventually sent shock waves through the stock market. The investor ended his interview with the observation, "Unlike 1869, today's stock market crash may lead to a widespread depression."

John shook his head at those words. "Farming has been rough enough this decade; this won't be good."

During the Great War, the U.S. federal government encouraged farmers to produce all they could to feed the soldiers and supply a war-torn Europe. Many farmers borrowed on government and private loans to buy modern equipment so they could farm more land. Before he left for the war, John pleaded with Berta not to borrow money. The wealth that John's family accumulated by moving from France to the U.S. meant they did not need to take out loans. This alone saved the Hoffman farm during the agricultural depression of the 1920s. As Europe rebounded after the war, American farmers continued to overproduce, and commodity prices fell. John, who held no loans, was unaffected. Many farmers who held sizeable loans were not as fortunate.

Farm bankruptcies swelled, and John wondered if this eventually had some effect on the stock market. Whether or not that was true, the crash was here.

Once worth $60 per acre, John suspected current land values had fallen to about $30 to $40 per acre. His land, some of the most productive in Indiana after it was reclaimed from the swamps, produced some of the best pastures around. John's father had purchased this farmland through the land-grant system, paying just $1 per acre after selling his land in France for $50 per acre. Lincoln signed the Land Grant Act to sell U.S. government-owned land in any state to raise money to build the Land Grant colleges. Lincoln's goal was to make higher education more available to the general public and allow people to move west and cultivate more land. Hoffman's benefited from this act. Owning good but inexpensive and debt-free land allowed them to survive the 1920s. The stock market crash worried John, though. He had no idea what atrocities loomed for farming in the 1930s. He didn't look forward to another decade of low prices for agricultural goods.

Chapter 12
The hard 1930s

A heatwave and dry weather hit the Hoffman farm in March of 1930. Fortunately, the soil's moisture levels were high when the year began, and deep-rooted alfalfa could survive one summer on the soil's moisture alone. Corn prices were still low after too many years of overproduction, so John could buy all the grain he needed for cattle supplement if necessary. This part of Indiana had been a swamp for multiple millennia, and the water levels helped the Hoffmans produce hay, oats, and corn. An extended drought of two years would challenge that. By the end of the 1930 growing season, the Boulton area had only received about sixty percent of its normal rainfall. Nevertheless, that was much better than other parts of the Ohio and Mississippi river valleys, which received only about one-third of their normal rainfall. John thought surely that summer would be the only dry year.

After the stock market crash, John not only used corn to supplement the cattle feed, he began to burn it in the kitchen stove to cook and heat the house. Corn was worthless on the market but cheaper and more accessible than gathering wood. Early that winter, before the heatwave started, it was cold, but they kept the house plenty warm because of the glut of corn.

Farm prices and a poor overall U.S. economy weren't the only troubles for the Hoffmans. That spring, a strange car drove up the lane. John, Karl, and Fritz had driven into town to get supplies. Berta didn't notice the car drive up to the house. She was accustomed to strange vehicles that drove up to the Jardin. John had created a space where up to ten cars could park so people could walk through the garden. The driver of this car ignored the garden and came directly to the house. When Berta heard the knock upon the door, it startled her. She glanced out the window and didn't recognize the car. The stranger knocked again. Slowly and hesitantly, Berta called out, "Who is it?" At the same time, she reached into the kitchen closet and pulled out a shotgun. She set it against the icebox.

"Sorry to disturb you, ma'am. I'm Reverend Duncan, at your service," the stranger shouted through the closed door.

Berta opened the door slightly and saw that the man wore a clergy collar. She opened the door wider and apologized, "Sorry, I didn't hear you drive up."

"Again, I'm sorry, but I'm here to help you."

Confused, Berta asked him inside. They sat alone at the kitchen table. Amelia had taken the two young boys outside. On this warm March day, temperatures had climbed well above average, so they strolled outside without jackets.

As the boys walked past, Reverend Duncan pointed at Samuel and asked, "Is that your youngest?"

"Yes, that is Samuel," Berta answered slowly, perplexed by the question. "Do you mind telling me what this is all about?"

A chubby man with a round face, Duncan smiled and replied, "Sure. As I said, I'm here to help you. Is your husband here? He might want to hear what I have to say."

"He's busy on the farm," she lied. Then her perplexed expression turned to perturbance. "Just say what your business is."

The reverend noticed Berta's change in mood, cleared his throat, and said, "I've been told by someone who carries your family in high regard that your youngest son was born out of wedlock." Berta stood up, shocked by his words. The reverend extended one hand and said, "Please let me explain. I come from a special part of our church that looks after things like this."

"Like what?" Berta demanded, growing very angry. Little did this preacher know that while Berta might appear frail, she was tough in mind and body.

"Well, people talk," he continued. "I'm afraid this issue will be with Samuel all his life if we don't do something." Berta took two steps toward the icebox. The reverend, sensing his time was short, blurted out, "If your family joins our church, I can arrange for a special dispensation for Samuel. You can be one happy family again." He saw the fire in her eyes but mumbled, "All that for a mere donation of $400 to the church to pay for the paperwork."

At that, Berta picked up the shotgun and said, "I don't know what church you are from, and I don't care. What I want is you off our farm right now." The reverend held both hands up and pleaded, "I'm only trying to help."

"You help yourself right back into that car and leave," she retorted. "Get out!"

Visibly shaken, Berta watched the reverend to make sure he drove away when Amelia and the boys returned to the house.

She and the boys ran to embrace her. "What happened?" Amelia asked

Slowly, Berta replied, "I don't know. That is the strangest thing that's ever happened to me. The man I ran off with the shotgun wanted us to buy..." She looked at Samuel. "He wanted us to buy into being one whole family again." Amelia pursed her lips. She understood. Berta burst into tears. After a moment, she looked up at Amelia, tears streaming down her face, and said, "I hope I didn't do the wrong thing, but he insulted me with his insinuations."

Months passed, but Berta kept the story of the pastor's visit from John, just as she had when she had been sexually harassed years earlier. The two incidents wound tight as a knot around her emotional core. Mentally, she attempted to rationalize her response to the reverend. She had never stopped loving John, and he never stopped loving her. She believed they were within their rights to conceive and birth Samuel. Somehow, the pastor's visit planted doubts in her mind.

During dinner one evening in May, Karl stood and took Amelia's hand and helped her rise, as he had done on Christmas Eve. He simply said, "We have an announcement." He looked at Amelia as if to say, You tell them.

She shook her head in disbelief that he wouldn't share the news with the family, then said, "We are expecting!"

Applause broke out, and John thought to himself, I'm going to be a grandpa?

Berta smiled. She knew Amelia was pregnant but hadn't said a word until they were ready to make the announcement. She reached out to Amelia and said, "I couldn't be happier that you will make me a grandma." Then she turned and struck John lightly on the shoulder. "You are going to be a grandpa!"

Fritz shook hands with Karl, who produced a handful of cigars. John, Fritz, and Berta gladly accepted them. Amelia passed because she never took to smoking anything, let alone a big, manly cigar, but Patrick accepted one, to everyone's surprise. The adults enjoyed a puff of the fine cigars that Karl had bought at a special Indianapolis cigar store. Suddenly, they heard coughing and turned to see Patrick, red-faced and choking. As he finally gained his composure, Samuel grabbed the cigar from him and took a deep puff from it. To the surprise of everyone, the youngest Hoffman didn't choke at all. He slowly exhaled and showed obvious pleasure in the treat. Patrick coughed again and said, "Well, I guess I am the baby of the family, AGAIN!"

Karl ruffled Patrick's hair and said, "On this." He held his cigar sideways, gave his brother a look, and added, "That's just all right. This is not for everyone."

Another month passed, Fritz always on the edge of his seat as cars visited the Jardin. When he could no longer stand the tension, he told Karl that he had to go back and visit the Blind Pig to see if there was any memory of him or his leaving the place. He finally explained to Karl that he had witnessed a murder and left Indianapolis the next morning.

"You've waited all this time to tell me?" Karl exclaimed.

"I was too embarrassed," he confessed. After a little more talk, Karl agreed to accompany Fritz on a visit to the Blind Pig.

The trip meant they'd be out late, so they got a hotel room and stayed the night. Karl considered the mission to be dangerous, so he insisted that just the two of them go, simply telling Amelia that Fritz needed to wrap up some unfinished business in Indy.

"Do you trust me?" he asked his wife. She didn't press for additional information, and on the afternoon that the brothers left, she instructed Karl, "Go help your brother. We all know that he's been uneasy for years." She kissed him before he walked away.

Karl pressed Fritz as they drove to Indianapolis. "I still can't believe you witnessed a murder and only now told me," he said.

Fritz rubbed his forehead and answered, "I didn't know what to say. Lizzy dumped me that day, and my boss was shot that night. I threw up in the bathroom after they left." He paused and thought about that night. "What amazed me the most was that they went about it so-matter-of-factly and then offered me a job. I've been so afraid they would track me down and kill me."

Karl shrugged and said, "Well, I'm glad they didn't. Who knows if they would have stopped with just you?" Fritz frowned, then Karl smacked him hard on the chest. "I'm just kidding. I hope."

The men checked into a hotel Fritz had chosen near the Blind Pig and cleaned up. Fritz, silent, sat on the edge of the bed.

 "Now what?" Karl asked.

"I don't know if I can go in there. They might recognize me."

"So, you want me to go in alone, without you?"

Fritz nodded.

"Geez, Fritz."

"They won't know you. You can go in and ask around."

"You want me to ask gangsters, 'Hey, do you remember my brother Fritz who witnessed you kill a man?'"

Fritz rolled his eyes and then stared at Karl. "You'll know what to do," he said.

Karl extended his hands and said, "Okay, but this will cost you liquor money." Karl entered the speakeasy at 9 p.m. He didn't have the appearance of a farmer; instead, he wore the sharp pin-striped suit from his wedding. There were about twenty patrons inside and a few, mostly women, watched Karl walk in. He approached the bar, pressed a hand down onto it, and asked the bartender for his best Kentucky bourbon. Karl hated that Fritz didn't come along, but he didn't let that stop him from enjoying the beverages. The bartender asked, "Are you sure you want the best? It's expensive."

Karl looked at him and said, "Hit me." He wondered if he was too melodramatic. A couple sat at the bar next to him and studied the stranger who had just ordered the expensive bourbon.

The bartender left and returned with a neat pour in a glass and said it would be $5.50. Karl slapped down $8 and told the bartender to keep it. He raised the glass to the couple next to him and said, "Cheers."

After two more of those, after all, it was Fritz's money, Karl called the bartender over and asked, "How long have you worked here?"

"Since 1925, why?"

"I visited here once back in 1925. There was an interesting bartender, his name was…" Karl snapped his fingers. "…his name was Fred. No Fritz. Fritz was his name! What happened to him?"

The bartender pointed at Karl and said, "I remember him. I was a customer when he waited for the bar. Friendly and knowledgeable guy."

"Yes, that sounds like him. What happened? It seemed as if he'd work here for the rest of his life."

The bartender replied, "Funny thing. You see, this place changed management overnight. I heard that Fritz bailed on the new management and skipped town."

"Did they ever wonder where he went?"

"I didn't get that impression. They acted like it was his loss. It was my gain, though, because they hired me the next day."

Karl smiled. "Hit me with one more of your best, and I'll be on my way." Karl drank it in one slug, smiled at the couple who nodded back at him and again left a good tip with Fritz's money.

He returned to the hotel room a little tipsy and told Fritz, "Come on. We are going to another place for a drink."

Fritz held out his hands as if to say, So, what did you find out?

"You've always had too high of an opinion of yourself," Karl said. "They remembered you but figured you skipped town." Karl shrugged. "Just like you did." He looked at Fritz and added, "You are off the hook. You don't have to worry about a mysterious car driving up to our place. Let's go somewhere else and have a drink. Do you have a recommendation?"

"Yes, if it's still there. It will be very removed from the Blind Pig."

Karl, who now felt the bourbon kick in, walked into the second bar and said loudly, "If you didn't vote for Herbert Hoover, you should leave now."

Fritz frowned and started to tell his brother to pipe down, but two men and a woman at the bar raised their glasses, "Hear, hear!" they said. One told the bartender, "I'll buy that man a drink."

Fritz glanced at Karl, who shrugged and said, "We are in Indiana. Everyone in here probably voted for Hoover." Karl walked toward the bar and accepted his free drink with a cheer to the man who bought it. He looked at the bartender, pointed his thumb at Fritz, and whispered, "I'll buy this poor, lonely

Democrat some moonshine." Then he slapped more of Fritz's money down on the bar. "Oh, and I'll also buy one for these three outstanding Republicans."

In December, Karl and Amelia welcomed Hollie Amelia Hoffman to the family. Born on the twelfth of the month, they felt it was close enough to Christmas to use the old English name that literally referred to the Holly tree. In a bleak 1930 winter, they enjoyed the warmth of a baby girl.

Berta was ecstatic when she and John visited. Finally, a baby girl in the Hoffman family, she thought.

Karl immediately brought more corn into their house to keep the furnace burning. He vowed his daughter would never experience the hardship of the early part of this century. The couple often stood over Hollie's cradle, Karl pulling Amelia close. Sometimes, she sang softly as the baby slept until they finally retired on a dark Indiana evening.

At dusk, a few weeks later, as the Hoffmans were again enjoying Christmas music on the radio, the announcer said, "Sorry to interrupt the music, but we have an update that might interest you. Our elf radioed us from the North Pole and reported that Santa has left the North Pole. He's headed south and will sweep back and forth across North America for the rest of the evening. We will keep you updated, but now back to Christmas music."

The Hoffman family applauded with delight. They would enjoy the traditional updates on Santa's progress from the Chicago radio station for decades.

Chapter 13

Frank and Singh Visit 1937

The heatwave and drought held their grip on the Boulton area until 1931. Though northwest Indiana received a little more rain than the Ohio Valley and much more than the Great Plains, The Hoffmans lost a few cows, and the soil in the garden cracked and hardened, necessitating that the Hoffmans buy more food from the grocery store in town until the heatwave subsided.

John heard that out west, some farmers had to choose between water for the livestock and water for the house. On what would have been a clear, sunny day, he saw dust thick in the air, blowing from west to east. The radio said it blew all the way from the Great Plains to the Atlantic. The sky above the Hoffman farm never darkened, but the dust created a brown ribbon flowing in the upper atmosphere. The radio also reported that the Mississippi River had fallen to its lowest level in recent history. Barges sat on sandbars in the riverbed exposed by the drought.

But by 1932, the weather began to improve, at least in the eastern Midwest, and the farm turned profitable again.

The world grew more modern each year. The Boone County Rural Electric Membership Corporation (REMC) became one of the first federally funded electric projects in the country, and the first in Indiana, on July 22, 1935. Later that year, farmers in Belton County received electricity. REMC ran electricity up to an eighth of a mile from the road for free. Hoffman's driveway extended beyond an eighth of a mile, so they paid for poles to be set and the wires to carry electricity up to the house. John didn't mind; he was happy the whole farm had power.

Karl and Amelia welcomed a son, Karl John Hoffman, into the family the following year. And then, in the spring of 1937, Frank returned to the Hoffman farm to see the completed Jardin. He also looked forward to sampling the Boulton moonshine again. On this trip, Frank brought with him, Lieutenant Abhishek Singh.

Karl met the two men at the train station. Frank reintroduced himself first with a hardy handshake. "It's great to see you again. You have filled out quite well. It must be the hard farm work, eh!" He poked Karl in the stomach and then gestured toward the lieutenant and said, "This is Abhishek Singh. But none of us westerners seem to say his first name correctly, so we just call in Singh."

Karl looked at the tall, dark man with a face full of beard and sad eyes. "Is that okay?" he asked.

"Of course, it is," Singh replied.

People at the train station looked upon the three, some shaking their heads and staring. Others tried to hide their glances and then turned away. After a few minutes, the men received their luggage, and the three proceeded to Karl's station wagon. Passersby glared at Singh, and people began to grumble about the man who accompanied Frank to Boulton.

Before they could climb into the station wagon, Karl said, "I heard that chocolate helps after traveling a long way." He led them to the local soda shop to buy the men a chocolate soda. "We can now get chocolate in a bottle. It's a modern world!" he added. As they prepared to leave the shop, the owner called Karl back to the counter.

"Does your papa know one of them is black?" he asked.

Karl, keenly aware of the racism in the area, said plainly, "Well, he's not black. I will admit that he's very dark-skinned." He looked at the man and added, "Sure, Dad knows he is dark-skinned. They served together in the war. That man was my father's lieutenant. Mister Singh is one of the most highly decorated British soldiers from the war."

"That may be, but he can't stay in the county overnight," the shop owner replied. "There are ordinances against that."

"Those are out-of-date ordinances," Karl retorted. "He is going to stay at my place."

The owner winced and said, "In the same house as with your wife, child, and baby?"

"That's right. Now, if you'll excuse me, I will take these men to our farm."

On the drive, Karl attempted to explain some of the issues that Singh might face while he was in the Boulton area. "It may be a long two weeks for you," he said.

"Nonsense, my friend has seen that before," Frank interjected.

"My son, racism is everywhere," Singh said solemnly.

"I wonder if you've ever seen it as bad as this," Karl said.

"I guess we will find out," Singh replied. "But I will tell you that it exists in many different forms in India. Race, castes, religion, social standing, gender..." he hesitated. "It's all different, but it's still about one group of people prejudiced against and ignorant about another group."

The men were silent for a minute, then Karl changed the subject. "See those poles and power lines?" Frank and Singh nodded, "We just got electricity two years ago."

"Well, I remember you had a telephone the last time I was here," Frank said, "Your old man never uses it to call me, though."

"You'll have to address that with him, but being from the last century, he's just old-fashioned about using a telephone," Karl laughed, "Kind of a dinosaur on that one." Then he realized the two men he was escorting were as old or older than John. "Sorry, no offense."

Singh said, "None taken. To have lived a long life after what we've been through is a blessing, not an embarrassment." Karl knew he was going to like Singh.

At the same time that the men were driving home, Samuel approached Amelia and indicated he didn't feel well, placing the palm of his right hand on his forehead. She immediately replaced his hand with hers and quickly determined that he had a mild fever. "Berta, come here," she called. Amelia motioned for Samuel to have a seat at the kitchen table. The fifteen-year-old sat quietly, watching Amelia go to the cabinet and find a thermometer. Berta walked into the room and saw that Samuel looked puny; his skin was pale. Patrick entered the kitchen and sat across from Samuel, ready to interpret, and Hollie walked into the room to see what the commotion was all about.

As they awaited the results on the thermometer, Hollie impatiently demanded, "Do mine! Do mine!"

"Just be still," Amelia said. "We'll need to clean the thermometer before we take your temperature." She put a finger on the tip of Hollie's nose, who giggled.

Berta pulled the thermometer out of Samuel's mouth and read it. "100.5 degrees," she said. "What's wrong, Sammy?" That was a name she rarely used except when she wanted to comfort her son. Samuel signaled to Patrick, who said, "He said he has a headache, and he feels very weak."

Berta sighed and said, "Well, young man, you need to lie down and get some rest. Patrick, can you grab a glass of water and follow me upstairs to the bedroom." Samuel pulled back the neatly folded bed covers, took a drink of water, and slipped into bed. He turned on one side, glancing back up at her as she sang to him. Within minutes, he was asleep.

The men arrived home at 6 p.m. Frank shouted immediately, "Stop! Stop this vehicle!" He jumped out before the car completely stopped and stumbled a bit. With both arms raised above his head, he surveyed the garden. "By damn, John, you've done it!" John, who was working in the barn when the three arrived, made his way down the lane to greet Frank and Singh.

Singh, less boisterous than Frank, quietly entered the garden. Amazed, he said to himself, it is just as we saw it under the German flares. Then he put an arm on Frank's shoulders and simply said, "Marvelous."

A minute later, John joined the two men, and Karl took the car up to the house.

"Well, I'll be," John mused. "Never thought I'd see the two of you here in Indiana." His mind struggled with images of the battlefield and rubble against this backdrop of lush pastures and the neatly organized Hoffman farm.

"When did you finish this?" Frank burst out.

"Well, it's never really finished," John answered. "It takes a lot of maintenance." He looked Frank in the eyes and said, "I'd say about three years after you left, it was complete, but as I said, never finished."

As the three strolled in the Jardin and admired the statues, shrubs, and roses blooming beautifully in the late-afternoon sun,

Frank asked, "Did you bring over the artist who made the sculptures for the original Jardin?"

Singh chortled, "You know he's been long dead, Frank." Frank shook his head, a little embarrassed that he'd lost track of time.

John answered, "We are blessed that Indianapolis has been actively building phenomenal fountains and war memorials. I hired a guy from there."

"Do I smell apple strudel?" Singh asked.

"You do!" John replied. "It's the peach-colored Lady of Shalott roses, just like in France, only they weren't blooming when we did our patrol through there. When I was younger, I had the advantage of seeing that Jardin in full bloom." Karl and Fritz joined the men in the garden while Berta and Patrick busied themselves and looked after Samuel. Amelia worked on supper, assisted by Hollie.

When they entered the house, laughing and speaking loudly of old times, Berta quickly came downstairs from Samuel's bedroom and held one finger to her lips. She extended her arms and motioned for the men to step back outside onto the porch. They complied without argument. Once they were outdoors, Berta said, "Gentlemen, I'm sorry to greet you this way, but our youngest boy has come down with a fever. I just got him to go back to sleep. May I ask you to keep the banter out here on the porch? You can watch the sunset while Amelia and I finish making supper."

Frank, comfortable with the family and fond of Berta, replied, "We don't mind at all. I hope the young man is not too sick."

"It's a low-grade fever. I hope it passes by morning. I want him to rest all he can," she answered.

Frank introduced Berta to Singh, who said, "I'm charmed for sure." Berta's face turned slightly red, and then she curtsied and returned to the kitchen.

The men settled into rockers and benches on the west side of the house. Fritz returned from the barn and produced moonshine, and Karl retrieved a fancy tray of glasses from the kitchen. Everyone but John filled a glass. Frank looked him over and said, "That's not the John that I know."

John frowned and replied, "There have been many changes, Frank. I won't bore you with the details tonight. Tell me, how was your trip across the pond?"

Singh piped in, "It was four days too long on a boat." The voyage actually took three and a half days, but Singh, who didn't care for boats, stretched the truth. He took a sip of the moonshine and uncharacteristically sneezed. "What is this?"

Frank said, "That will put hair on your chest," and burst into laughter.

John apologized, "We can get you a bourbon or wine if you prefer. This is the drink the boys have come to appreciate."

"Me, too," Frank said as he held his glass up to refract the sunlight.

Singh shook his head. "That won't be necessary," he protested. I sometimes sneeze on the first sip of strong alcohol if I haven't had a potent drink in a while. Its strength reminds me of Indian feni, especially the kind made from toddy palm." Singh took a second sip. "Not as sweet as feni; that's what especially took me by surprise."

Changing the subject, Fritz asked, "What's it like to be on a great big boat and cross the ocean?"

"Smelly, especially around the lavatories and in the not well-vented hallways," Singh offered.

"But you had to be amazed by the view of New York, especially the Statue of Liberty," Frank insisted.

Singh nodded but said, "The view from the train from New York to Indianapolis was far more attractive. The contrast of the green mountains with white steam puffing toward the sky, the rolling hills, and then the open farm country. This country truly is the amazing machine that fed Europe during the war."

Frank raised his glass again and said, "Hear, hear!"

After dinner, Berta, Patrick, and Samuel stayed back as the rest of the Hoffmans and their two guests moved to Karl and Amelia's house to continue their visit. They discussed the visitors' long trip, the war, and racism.

Late that night, John stood and announced, "It's time for me to go home. Frank, we have a room for you at our house. Singh, Amelia has you set up in the guest bedroom."

Frank patted the couch he sat on. "I don't want to disturb the boy," he said graciously. If it's okay with Karl and Amelia, I'll sleep fine here."

The couple looked at each other, and Amelia said pleasantly, "I'll get you some blankets."

Chapter 14
Samuel Struggles 1937

The next day, Samuel's temperature climbed to 101.2 degrees.
Berta readied to take him to see Doc. John offered to take the
two to town and asked Frank and Singh if they wanted to visit
Boulton a little more. The men agreed that they would, and
Patrick road along in case he needed to interpret Samuel's
signals.

A crowd was gathered at the end of the driveway, and John
assumed they'd come to view the Jardin. But as John passed,
with Berta and Samuel in the front seat and Frank and Singh in
the back, the people shook their fists. Singh chose not to look at
them and stared straight ahead. Frank looked them up and down

and said, "They are on your property, John. You should have them arrested."

As they pulled onto the road, John said, "They know exactly what they are doing. They are standing on the roadway property easement."

The drive into town was quiet, each absorbed in their own thoughts. John, concerned about Samuel, also thought about the disturbing people at the end of the drive.

Singh noticed a bright, swollen spot on Samuel's neck. He pointed at it and asked, "What is that?"

Berta put her hand on the back of Samuel's head and gently leaned him forward to examine the spot. "I don't know," she said. It wasn't red this morning when I checked on him."

"It looks like a tick bite," Singh said.

"We'll have Doc look at it," Berta said.

John parked the vehicle in front of Doc's house. Berta and Patrick took Samuel through to the office while the men proceeded to walk three blocks to the downtown area. People stared at Singh, and the party was greeted with frowns when they entered the shops. Sheriff Bucklyn met them at the door when they exited the third store. Looking directly at John, he said, "We have to talk."

John looked at him and simply said, "Go ahead. Talk."

"I'd rather talk in my office," the sheriff said. A few people gathered in the street, eager to hear what was being said.

John looked down at the sidewalk and then directly into the sheriff's eyes. "Whatever you have to say, you can say it here," he stated.

Slowly, the sheriff drawled, "You know the ordinances."

"I do."

"There's a fifty-dollar bond for a..." the sheriff hesitated. "...for a black man to stay in this county."

Without hesitation, John pulled out his wallet and handed the sheriff fifty dollars. "This should cover the bond."

Singh also reached into his pocket and pulled out fifty dollars. He said, "John, I'll pay it."

Without looking at Singh, Sheriff Bucklyn said coldly, "I'll only take the money from John."

"Put your money away, Singh. This one is on me," John said.

"We will talk about this later," Singh answered calmly.

As the sheriff walked away, he realized that people had watched him take money from John. Turning back, he said, "I will mail you a receipt." John knew he would not.

The men walked silently back to the vehicle, and John invited them to have a seat in the station wagon as he went into Doc's office. Berta stood and listened to Doc say, "This definitely is a tick bite. I must think the fever will break in a few days. In the meantime, give him some aspirin to help break the fever."

"Thank you, Doc," Berta said. Then she turned to look at John and asked, "What's the matter with you?"

"I'll explain later," he answered.

Berta turned back to Doc. "Is that everything?" she asked. "Is this a normal eight-dollar visit?" The doctor nodded, and Berta pulled the money out of her purse.

Once they all climbed into the car, John explained what had just occurred with the sheriff. Berta covered Samuel's ears and said, "That doesn't surprise me from that A-hole."

John glanced at Berta in surprise. "You never talk like that! I sense a great deal of animosity toward the sheriff."

Berta thought for a moment and told them about the day she and Amelia wanted to report the sexual aggression toward her.

She spoke carefully because the two boys were in the car. "We were separated then, so the sheriff blamed it on me," she concluded.

Frank patted Singh on the shoulder. "It appears you weren't the only one offended by this constable, my dear friend," he said.

John changed the subject, and the group participated in small talk the rest of the way home. It wasn't until after dinner that John broached the issue again. His family gathered 'round the table. "You see, racism is extreme here. We have had state laws that didn't allow blacks to start a business here in Indiana. It's preposterous. Until a few years ago, the Ku Klux Klan..." John stopped as he realized he should explain to Frank and Singh, "...a group of so-called white supremacists attempted to keep the state 100 percent white and Protestant. It's a job they largely would have succeeded in if it weren't for the huge flow of German Catholics and the scandal related to the Indiana leader of the KKK. That scandal, thankfully, forced the downfall of an organized KKK in Indiana. Don't get me wrong. We are proud Protestants. Those in the KKK were just a few compared to the rest of us."

"I see," Singh sighed. Again, he produced fifty dollars to reimburse John.

John waved it off. "Please, let me pay it as an apology because you were confronted so."

"Well, it was you who was confronted," Singh said. John nodded but did not take the money.

The town gossips learned of Samuel's fever from people in Doc Bane's waiting room. That night men gathered for the usual card games at The Station. Moonshine spoke about it first. "I hear that the black man brought the fever over on the boat with him. It was on a tick." The men around the table groaned.

One person said, "Fig'ers."

The men grumbled about the visitor for the rest of the night. Finally, Moonshine said, "We have to do something about it tomorrow."

The next day, Moonshine and the others showed up in the parking area of the Jardin, but Berta noticed that their attention focused on the house and not the garden. Meanwhile, several people drove by Karl and Amelia's house, where Singh slept during his visit.

Fritz and Patrick tended to the farm chores while those inside watched the gathering from the windows, which had grown to twenty. Berta said, "John, this isn't good."

"This could get serious," he answered.

"What? Why do you say that?" Frank asked.

"Just six years ago, a mob lynched two black men down the road in Moraine," John said quietly.

Karl interrupted, stating, "Those men killed a white man and raped his fiancée. That was proven in court."

"Still, it was an example of a mob out of control," John said. "Karl, call the sheriff."

"The sheriff won't do anything," Berta said with exasperation.

John gave her a look and said, "We bought the sheriff, remember?"

Just then, Samuel walked into the room as everyone stood staring out of the windows. He signaled as if Patrick was there. Berta tried to interpret, but without Patrick, no one knew what he was trying to say. Suddenly, Samuel fell to the floor. Berta rushed to him and drew him into her arms. His arms fell limp at his sides.

"John, we have to take him to Riley," Berta cried.

"You two must come with us. I'm not leaving you here," John said to Frank and Singh.

Two hours later, now with Patrick along, they approached the hospital emergency room entrance. "You two go," Frank directed John and Berta. "I'll find a place to park. We'll find you." John carried Samuel as they bolted from the car.

The emergency room staff laid Samuel on a gurney and rushed him into an available room. John stayed at the admitting desk, and Berta followed the bed as a nurse rolled it into place. The nurse began to check Samuel, who breathed normally but was unconscious. Moments later, a doctor arrived and looked Samuel over. Berta pointed at the tick bite.

"This isn't good," the doctor said calmly. Then he turned to the nurse and said, "Try to take care of his fever."

John entered the room and stood across the bed from Berta. Returning his attention to John and Berta, the doctor said, "I might have an idea what this is. I need to make some telephone calls to gather other opinions. I'll get right back with you."

"What is it?" Berta cried as the doctor started to walk away.

"I can't say until I'm sure," he answered, and Berta knew by his tone it was serious.

"We're going to move him into his own room. Follow me," The nurse directed Berta.

Back in Boulton, the sheriff showed up on the Hoffman farm, where the crowd had gathered to protest Singh's presence in the county. They waited for his return.

The pudgy sheriff rolled out of his police car and addressed the crowd. "John Hoffman has the right to keep a colored guest. He has paid the bond. I want you all to go home and leave the Hoffmans alone."

Karl and Fritz approached the end of the driveway to watch.

One of the protestors shouted, "It ain't right. There are laws."

The sheriff continued, "As I said, John addressed the law. He paid the bond." He would not admit that he had pocketed the fifty dollars.

"He brought that fever that's made the boy sick," the protestor continued. "We might all get it."

The sheriff shook his head and replied, "I talked to Doc about it. It turns out that Samuel was infected by a tick before the black man arrived on the farm."

In Indianapolis, a short while later, the hospital doctor returned to Samuel's room. Berta and John stood on each side of the bed where their son still lay unconscious. "How long has he had the fever?" the doctor asked.

"We think five days," Berta answered. "Tell us what is going on!"

"I see. There's no good way to explain this, so I will be straightforward. I am very sure the young man has a rare form of encephalitis." John and Berta both showed puzzled faces. "It appears…" the doctor started, then hesitated. Clearing his throat, he said, "We will take some blood and do some tests, but from what others have told me, it appears to be fatal."

Berta gasped. John reached across the bed to grasp her hand and attempted to hold her steady. The two stood in shock.

"Has anyone else in the family had a fever?" the doctor inquired.

Berta shook her head.

"Have you been anywhere in a wooded area recently?" the doctor asked.

"The older boys took Samuel to the Portland Arch, south of Attica, just days before the fever started," John offered.

"I see," the doctor said again. "This disease is so rare here in the United States that we don't know much about it except that it is fatal in a small percentage of cases and usually within seven days of when the fever starts. I'm afraid that this is one of those times. I assure you that we will do everything possible to keep him comfortable."

Berta said to John, "We should have brought him here when he first showed a temperature."

"What was the first temperature you took?" the doctor asked.

"It was 100.5. We took him to see the doctor in Boulton the next day when it climbed to 102 degrees."

The doctor put his hands on the bed rail. "I see," he said. "I must explain that there is no way your town's doctor would know about this. I only just read about it in the medical journal this week. If you will excuse me, I want to do more research. Others will be in to take some blood."

An hour later, the doctor returned. He scratched the back of his neck as he walked into the room. "From what I've been told, his brain has swollen severely," the doctor said. "That's why he's unconscious. It looks as if we can do an X-ray. But that involves surgery. It's an experimental practice where we blow air into the brain cavity."

John interrupted, "You what? What good will that do?"

"Frankly, it will only let us see the brain better," the doctor answered. He walked over and touched the unconscious child's arm. "And, sometimes the surgery itself can be fatal. I'm afraid as weak as he is, it might be."

"John, no!" Berta cried, tears slipping from her eyes.

"Then there is no question that we don't want you to do it," John said.

"I see. I agree, but I just wanted to offer an alternative," the doctor replied.

The doctor left but returned to the room periodically to check on them. Eventually, John went to the waiting room and explained the situation to Frank and Singh. "Do you mind staying the night in a hotel?" he asked his friends. "I'd hate to send you back to the farm with that crowd." John had hoped the sheriff had done his job, but he didn't want to risk it. "I recommend the hotel just across the street. It also has an excellent restaurant. Did you bring money?"

Frank grimaced at the thought of John handing them money and said, "Don't worry about us." He motioned toward the exit with his head and said, "Let's go find a room, Singh."

Overnight, John and Berta watched Samuel turn paler and paler. They each kept a hand on Samuel and an arm around each other. Patrick fell asleep in a chair.

At midnight, Berta said, "There's something I have to tell you." John looked at her and indicated he would listen to whatever she had to say. She continued, "Eight years ago, when you, Karl, and Fritz were in town getting supplies one day, a reverend visited from a church I've never heard of before. He wanted us to pay money for a dispensation for Samuel."

"What?" John asked in a forceful whisper.

"He said we could be one whole family again. I chased him off with a shotgun, John. I didn't know what else to do."

"You did the right thing," John said calmly and hugged her tightly in his arms.

"But what if this..." she touched Samuel. "What if this is because I didn't make it right with God?"

"Don't worry about that," John said emphatically. "If that reverend wanted to make it right with God, he would have given the dispensation for free." The two grew silent and returned their attention to Samuel. His breaths became shallow. The night nurse assured them that they were doing all they could to keep their son comfortable. At 3:15 a.m., Samuel shuddered, and John and Berta watched him take his last breath.

Four days later, the Hoffmans and their two guests stood in a quiet cemetery outside Boulton and laid their son to rest. John and Berta's son, who had led such a challenging life, was gone way too soon. A local pastor offered a final prayer, and Fritz stormed away after the service. John watched him go, then looked inquiringly at Karl as if to say, 'Why?'

Karl extended a hand and said, "Stay here. I think I understand. I'll go talk with him."

Fritz bent over behind the car and vomited. Karl walked up behind him and gave him a moment before he asked, "Do you want to talk about it?"

Fritz mustered, "I should have been closer to Samuel. I'll never get that chance again."

Karl leaned against the car and folded his arms across his chest. He thought for a moment and said, "Samuel knew we loved him, and that includes you."

As the two older brothers talked, Patrick knelt at Samuel's coffin, and hand signaled as if Samuel could see him. Berta watched, tugging on John's sleeve to get his attention. She spoke softly to her husband, repeating the numbers that Patrick signaled. She said, "Twelve, fifteen, twenty-two, five. That's the word love." Oblivious to his parents, Patrick stopped for a second and then signaled again. Berta counted the numbers aloud, "One, twelve, twenty-three, one, twenty-five, eighteen. Always." She finally figured out the numbering system.

Berta pushed her face into John's shoulder and cried softly. John, on the other hand, thought about Samuel's death like a father hardened by war. He wished that wasn't the case. He glanced at cemetery markers, two for his father and his wife and two for his grandfather and wife. They represented two generations of Hoffmans in the new world. John mused to himself, 'Samuel represents the fourth generation. How long will it be before I join them?'

At the end of the week, as scheduled, Frank and Singh left Boulton. As they stood at the train station with John and said their goodbyes, John grew silent. Frank, reading John's thoughts, stopped him and said, "I don't have any children. Four years of war took care of that. I cannot imagine what you must be going through right now. So, don't apologize to me about the week."

Then he embraced John and said, "Just promise me that you and your lovely bride will visit me in London some time." John nodded silently.

Chapter 15

WWII 1942

Life on the Hoffman farm changed after Samuel's death. John focused on the Jardin a little more, and Fritz bought a piece of land. Though he still farmed with John and Karl, Fritz now spent less time with the family since he lived in his own house and also farmed his own land. All of this caused Fritz to distance himself from the rest of the Hoffmans. Berta expressed her concern to John, who simply said, "He'll find himself. Give him time."

Fritz had found love with Kathryn, an attractive woman from Lafayette, and the two had a baby boy, Jonathan. The day he was born, Fritz loved his new son and, for the first time, found what John may have felt toward his boys. He thought about playing with Jonathan as he grew. In jest, he said to Kathryn, "Now we have someone who will be able to help on the farm in a few years." Somehow having a baby brought some resolution to Fritz for missing out on years with Samuel.

Kathryn, like Fritz, enjoyed moonshine. She was more comfortable spending time with Fritz and his friends than with Berta and Amelia. They didn't mind. They attributed this to the couple's new and young love.

Berta felt the presence of the missing child in the home every day and wondered if that would ever cease.

Patrick, who was no longer needed as a caregiver and interpreter for Samuel, spent more time helping out on the farm. Every once in a while, Berta saw Patrick's hand signals as if he were talking to Samuel.

Four years after Samuel passed away, on December 8, 1942, the day after Japan bombed Pearl Harbor, all members of congress but one voted for the U.S. to enter World War Two. Patrick readied himself to join the fight. John protested but knew there was nothing he could do to stop Patrick, who was now twenty-two.

"War like this is a living hell if you are lucky enough to live through it," John said solemnly.

"I understand from the stories you've told," Patrick said. "But I must go. I have to do my share." Patriotic messages flooded the newspapers and the radio. The sense of duty moved everyone, men and women alike, to join the war effort. Patrick was no different. Like Fritz, he read the newspapers and deathly feared the expansion of the German Reich and its ideas. He also wanted to follow in the footsteps of his father, who had helped save Europe from German aggression.

At the armory in Lafayette, the army used a bus to pick up volunteers and send them to Kokomo to be processed. The Hoffmans watched as Patrick climbed onto the bus, and it drove away. Two days later, Berta stood in the kitchen when the telephone rang. When she picked it up, she heard Patrick on the phone. "I'm coming back home," he said. "They tell me I'll be at the Lafayette armory by 3 p.m. tomorrow. Can you come pick me up?"

Puzzled, Berta asked, "Are they waiting to ship you off to training?"

She was met by silence.

"Are you there, Patrick?"

"Yes. I'll explain tomorrow," he answered and then hung up.

John and Berta waited for Patrick at the Lafayette armory the following day. Though they were uncertain about why their son was returning, they sensed this was not good news. The army whisked young men off to training camp within hours of processing and a medical checkup. John suspected the reason but chose not to voice his thoughts until they could talk to their son.

Patrick stepped off the bus with two other men. The mood was vastly different from when Patrick left on the bus with twenty boisterous, testosterone-filled men just a couple of days earlier. Grimly, he greeted his parents.

"Well, tell us what happened," John demanded impatiently.

"They said I could not go because of a medical condition," Patrick complained. "They called me a 4-F'er and sent me packing for home. It really was embarrassing." Patrick cast his eyes toward the ground, unable to look at his parents.

"What's the condition?" Berta asked, her voice filled with concern.

"They said I have a weak heart." He looked back at the bus and thought about the medical examination. "They said I may not live ten more years."

Berta gasped. "They have to be wrong. You're so healthy."

"Stinks, doesn't it?" Patrick offered. "Maybe I'll get to see Samuel sooner than I thought."

"We're going to see about that," Berta announced firmly. "We'll take you to a heart specialist in Lafayette."

The next day Berta lined up an appointment for Patrick with a heart specialist that Doc had suggested. The appointment was two weeks away, the soonest she could get him in. She, John, and Patrick sat at the kitchen table and talked about it. She patted Patrick on the arm and said, "You look very somber. Tell us your thoughts."

"When I heard that I probably won't live another ten years, it hit me like a ton of bricks," Patrick said. He hesitated, then continued. "I thought about our family and the farm. Especially Samuel, who had so much to offer, but we were robbed of that. Samuel was robbed. Now it looks like I will be." Patrick had never been one to cry, which didn't change on this day. For him, life was a matter of fact. There were no shadows, only light, even in this dire situation.

The conversation was punctuated by the sound of two car doors slamming. They thought it was probably Fritz and his crew. They sat quietly for a moment. Sure enough, Fritz walked in with Kathryn, who was carrying Jonathan.

"Whoa! What are you doing here, Patrick?" Fritz exclaimed.

"The army won't have me," Patrick answered. "They say I'm not strong enough for the exertion of war."

"What? You should have kicked that medic's ass. That would have shown him."

John piped in, "They say he has a bad ticker." Fritz and Kathryn joined the others at the table. Kathryn handed the baby off to Berta.

Patrick nodded. Instead of talking about himself, Patrick then turned the conversation to Samuel. "He and I had a secret," Patrick said.

"The code? We know about that," Fritz replied.

"No, it's something he and I never told anyone," Patrick said. He stood and went to the icebox and pulled out a pitcher of lemonade, raising it to see if anyone else wanted a glass. Kathryn jumped up as several nodded and retrieved glasses from the cabinet. Karl, who had learned about Patrick's rejection that morning, walked in at that moment.

"You're just in time," John announced. "Patrick is going to tell us a secret."

The family members took turns filling their glasses. As John emptied the pitcher, Berta said, "Looks like I'll have to make more. But first, let's hear about this secret."

Patrick gulped some lemonade and said, "Samuel saw things all of his life...or, at least when he was old enough to remember."

"Like ghosts?" Fritz interrupted.

"No. Things, people, animals, shapes. They came randomly and at any time. They always appeared in his peripheral vision first. It always startled him. He often jumped."

Karl shook his head and mused, "That would explain that. I saw him jump often. I wish I had asked him about it."

Patrick nodded, then went on. "I did, and he told me. When he turned to look at what he thought he saw, it disappeared. He and I thought it was because of being hit on the head as a baby. It turned out to be a retina problem. I looked it up. He suffered from what they call metamorphopsia. We found out that the lines in his peripheral vision were curved, which caused him to see things that weren't there."

Berta's jaw dropped, and then she exclaimed, "Why didn't you tell us about this?"

"Samuel made me promise not to tell anyone. He wanted it to be our secret," Patrick answered.

Fritz put both hands on the table and stated, "You two were incredible."

"Why are you telling us now?" Karl asked.

Patrick gave him a look and said, "I think that's pretty obvious. He isn't here. I believe that relieves me of the promise. We researched it thoroughly, and there is no treatment for it."

Embarrassed, Karl simply looked down at the table. After that, the family sipped the lemonade in silence but moved out to the porch to watch the sunset.

For Berta, the two-week wait to see the heart specialist drew out like an eternity. Patrick functioned as normal and helped with the farm chores. Berta sometimes wondered if he wanted to say more about it, but she didn't press the issue.

On the morning of the appointment, the entire family gathered for breakfast prepared by Amelia. As they finished, Berta said, "I want us to hold hands and say a prayer for Patrick's appointment today." Quickly, everyone took the hand of the person next to them, and they bowed their heads. Berta said a few words and finished, "I pray, Lord, that Patrick receives better news today than he did from the army medic."

"Amen," Fritz said. He looked Patrick in the eye and added, "I have always worn my heart on my sleeve, spilled out my problems, and poured them on others. You have always looked out for others and their problems. Why is that?"

Patrick sat silently for a moment. Finally, he answered, "I remember when Samuel's accident happened. I got scared and sat under the kitchen table." Karl nodded at the memory. Patrick pointed at Karl and continued, "Then this big lug sat next to the table and reached for me. I resisted. But I'll never forget what he said: 'He will be all right. We will all be all right.'"

Karl's mouth dropped open. He had forgotten that conversation.

Patrick went on, "Ever since then; I've lived by Karl's words. Everything will be all right."

As Amelia cleared the table, she patted Karl on the head. "There you go, you big dummy. Sometimes you do offer pearls of wisdom."

Fritz laughed, then directed his attention to Patrick and their parents. "Well, you guys should get on in a jiffy!" The three looked inquisitively at Fritz, who replied, "It's a new word that means quickly."

Berta rolled her eyes and said, "Thanks for that!"

Karl and Kathryn helped clear the table as Berta, John, and Patrick prepared to leave for the doctor's appointment in Lafayette.

Patrick took a long look at the Jardin as they drove past it. "You sure have made a beautiful site here, Pop. You should be proud," he said.

John looked at his son in the rearview mirror and replied, "Your mother and I have made even more beautiful things."

At the doctor's office, John and Berta waited more pensively than Patrick. When the doctor called Patrick into his office, he surprised John and Berta by asking if they wanted to join them. They jumped at the chance and followed Patrick.

Inside, he instructed Patrick to sit on the examination table and remove his shirt. "So, you've been told you have a weak heart. Let's take a listen, shall we?"

As the doctor grabbed his stethoscope, Berta exclaimed, "He was told he might not have ten years to live."

The doctor looked at Berta and then at Patrick. "Who told you this?"

"The army medics that checked me when I volunteered to join."

The doctor put the stethoscope in his ears and on Patrick's chest." Within a minute, he shook his head.

"Is it as bad as they say?" Patrick asked.

The doctor set the stethoscope down and smiled. "Son, those army medics don't know what they are talking about," he said. "In the medical community, we are just learning about what you have." Patrick, Berta, and John hung on to the doctor's words. "What you have is a heart murmur. That's it."

"What's that?" Patrick asked.

"As we understand it, it's about how your blood flows in your heart," the doctor explained. "The army medic heard it, but he has no idea what it is." He patted Patrick on the shoulder and said, "You will likely live a normal and healthy life. You should be thankful the army medic heard that. He saved you from this godforsaken war."

John smiled when he heard those words.

The doctor continued, "It's too early for me to tell you what kind of blood flow issue you have in the heart, but I'm sure that it isn't deadly. We will do more tests to see if you need any medicine. I'm sure that you don't need any surgery. Any questions?"

"So, I can do farm work?" Patrick asked.

"As much as you want." The doctor turned to John and said, "Don't let this condition get him out of any work." He returned his attention to Patrick and said, "You can put your shirt back on. Stop at the front desk. I'm suggesting a different specialist who can more thoroughly find out exactly what is going on. They will set up an appointment."

As soon as they were outside, Berta exclaimed, "I want to go to McCord Candies and get treats for everyone!"

Chapter 16

The Sunset 1951

Changes continued to abound on the Hoffman farm. A television console, complete with a record player and radio, sat in the spot the old radio once occupied. Tupperware and a toaster oven were mainstays in the Hoffman kitchens, and the grandchildren, five of them now, played with Slinkies and Silly Putty. Patrick had married a woman named Katie, and they had two sons, Mike and Paul. They drove a brand-new Jeep, a vehicle that had quickly grown popular after helping the Allies win the second world war. The world was changing quickly, and sometimes John wondered if that was a good thing.

John also regretted that he had built the Jardin so close to the house because it was now drawing crowds of visitors, who often told him they'd driven from Chicago or Indianapolis to see it. One tourist from Chicago told John, "A friend told us it was worth the drive just to see it, but also to hear the story behind it." John obliged and told the story over and over again.

The U.S. economy boomed after the war. More roads were built, and the county paved the road that ran past the Jardin. John wondered if the garden's tourism was the reason for that.

The farm also prospered, and Karl, Fritz, and Patrick acquired more land. The Hoffmans became a powerhouse in cattle breeding and marketing. Several tornados scratched their way across Indiana, but luckily the Hoffman farm and the Jardin remained unscathed.

The Hoffman brothers' political views had clashed during the war. They often caused heated discussions between the two older brothers and their wives, who were staunch Republicans, and Patrick and Katie, who heavily leaned Democrat. The elder couples cheered the fact that the United States had become a world power, whereas the younger couple was angry that the United States had become the default world police. Over time the bickering stopped. John figured it was partly because his sons and their wives were busy with parenthood. And they were all in agreement about the new civil rights movement that was gaining attention. All of the Hoffmans were certain that the segregation of the past must end.

One sunny, late Sunday afternoon in 1951, after the chores were done, the Hoffman family sat in the Jardin and rested. Berta, Amelia, Kathryn, Katie, and Hollie sang folk songs with occasional input from Karl, Fritz, and Patrick. The younger grandchildren played hide and seek in the Jardin's shrubbery. John contentedly took in the scene and wondered if the children of Bailleul had played the game in the Jardin in their town.

As the family enjoyed peaches and ice cream on that afternoon, they also told stories and recalled fond memories. Patrick was quieter than usual, and Berta asked about it.

He looked up, smiled, and said, "I have another Samuel secret that I never told anyone."

Berta, who delighted in her son's smile, nudged him. "Go ahead."

"This involved the spoiled oats," Patrick started. John laughed because he knew where this was going. Puzzled, Patrick continued, "When Samuel was six, he picked some peaches and placed them deep into a wagon of oats." He explained to the younger Hoffman's that oats in a wagon under the lean-to always stayed cool, and putting the peaches in the oats helped cool the just-picked fruit that had been warmed by the sun. "That time, Samuel forgot about the peaches. Days later, when he finally remembered them, the peaches had rotted and began to spoil the oats. He asked me what to do about it. We got buckets and hauled those spoiled oats out to the pasture behind the barn. We never told anyone."

Again, John laughed out loud, which surprised everyone, especially Patrick. John laughed again. "You didn't have to tell me," he said. "I found the spoiled oats and peaches that were in plain sight behind the barn."

Patrick mused and then said, "And to think that Samuel and I thought for days that we'd get our butts kicked."

John shook his head and asked, "When have I ever done that?"

"John has a good point," Berta said, supporting her husband.

"Yeah, Daddy-O never did that," Fritz chimed in, unable to resist using a new term he'd just read in the Chicago City Times that morning. "Daddy-O is anyone who is a cool dude, not just a dad, but I think it applies here," he added as he watched his wife mull over the new expression.

Jason Schmidt from the Canton Foundation in Indianapolis traveled back to the Jardin several times. Sometimes he just wanted to get out of the city, and sometimes he attempted to talk John into opening the Jardin to the foundation. He had stopped by just the day before and made another offer. In his sixties, John began to consider those options, though he'd not shared that with his family.

Karl smiled and looked across the land, taking in the cars driving along the road with rubberneckers viewing the Jardin. "Dad, maybe you should take the foundation's offer and let them manage it," he said.

"Over my dead body," John quickly retorted. His answer surprised no one.

Fritz looked at his brothers and said, "Well, in that case, I think we need to adjust."

"Adjust what?" John asked.

Patrick jumped in, "I think where Fritz is going is that we step up and help maintain this, this Jardin," he said.

"Yeah," Fritz said. "I think us boys need to participate more in the beauty of your invention."

Berta smiled, knowing this conversation pleased her husband. Yet, the Jardin had been his dream. She remembered him calling it his penance for destroying the original in France. "The original was destroyed more than thirty years ago, John," she said. "I believe it is time to share the responsibility of maintenance."

The family sat quietly as they waited for John's response. Eventually, he offered, "And how would that work out?"

Karl didn't hesitate. "First, we allow people to enjoy it more. We take down the no trespassing signs you put up during the war to keep people from trampling the paths and picking roses." John frowned, but Karl continued, "It's greedy of us not to share this."

Patrick added, "It will take some hard work, but that's why us boys need to change. We can do more to help you manage this." He pointed to the southeast section of the Jardin, closest to the road. "We can build a small station there with plaques and signs that explain the history of the original Jardin, along with the rules." He hesitated, then said, "We can explain how it was destroyed in the first war." John winced, but Patrick went on. "And, we can explain how and why you rebuilt it. People see the statues, but they have no idea how they illustrate the great respect the original artists had for nature, rain, and fertility."

John seemed to relax.

Fritz had been listening quietly, and his face had become tense. "That will take a lot for us to manage," he said.

Berta, who knew how Patrick thought, said, "Let Patrick finish."

The grandkids had finished their peaches and ice cream and returned to their hide and seek game. Patrick pointed to them and said, "They will help, and the entire family will get involved."

John still wasn't ready to give up sole responsibility. Berta softly put her hand on John's knee and said, "Keep listening. Patrick is always practical."

Patrick nodded at his mother, then said, "We could go with Karl's idea and build a booth where people drive into the parking lot. We could install a dropbox where people can donate; we could suggest one dollar per car."

John shook his head and said, "We don't need money." His Indianapolis investments had quadrupled since the war ended.

Karl interjected, "Patrick, you're always the one to put wheels on a plan. But I want to take your idea one step further. We could contract with the local Future Farmers of America and the 4-H programs to provide the booth workers. The money would be a donation to those groups."

John warmed to the philanthropic aspect. So, Karl continued, "Of course, that would be after we put that money into a fund to replenish the Jardin. All we would have to do is manage the kids in the booth. You could interact with the visitors as you please."

Fritz, coming around to the idea, said, "Of course, we'd need to shut it down when it freezes and on rainy days. That's where I come in. I can work with the news media and speak at gatherings about what we want to accomplish."

"Yes, sure," Karl quipped. "Find a way to get out of the hard work." Fritz knew he was joking. They sat silently again for a few minutes, and then John said, "So your plan is that we don't hand this over to a big city foundation, but we do what they want to do and manage it ourselves."

Patrick nodded and said, "For as long as we can."

The boys put their hands in the air as if in a pledge and repeated Patrick's words: "For as long as we can."

John looked at Berta and said, "Maybe with the whole family involved in the Jardin, you and I can make the trip to London and visit Frank."

"I'd like that very much," Berta replied.

Fritz looked over the Jardin and watched the Gelbvieh cattle graze in the pasture. Behind them, the sun sat with bold white and red spikes piercing the sky. He spoke softly: "You know, we've been through wars, depression, gangsters, hard times, and great times, like right now. You know, we are the sunset kings."

-end-

Coming in 2022

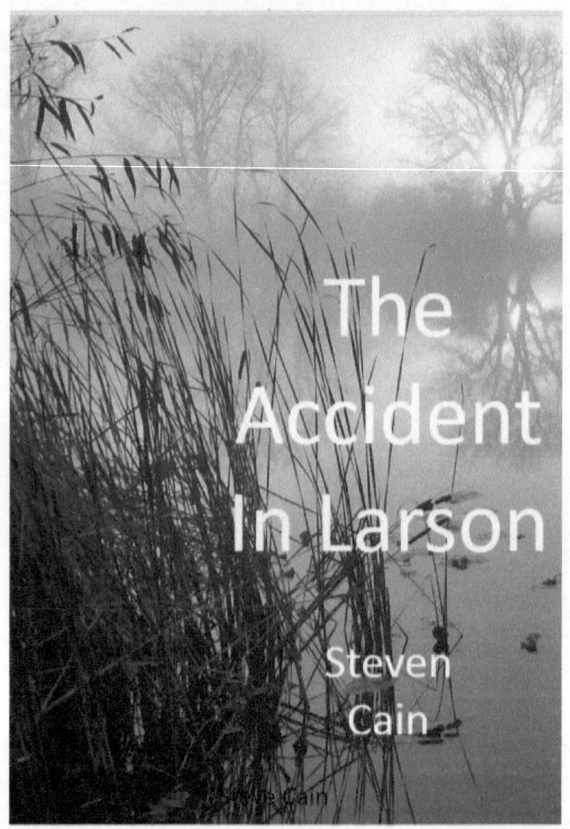

After a mysterious car crash, a young man struggles to rebuild his life in rural Indiana.